Solid Mechanics and Its Applications

Founding Editor

G. M. L. Gladwell

Volume 268

Series Editors

J. R. Barber, Department of Mechanical Engineering, University of Michigan, Ann Arbor, MI, USA

Anders Klarbring, Mechanical Engineering, Linköping University, Linköping, Sweden

The fundamental questions arising in mechanics are: Why?, How?, and How much? The aim of this series is to provide lucid accounts written by authoritative researchers giving vision and insight in answering these questions on the subject of mechanics as it relates to solids. The scope of the series covers the entire spectrum of solid mechanics. Thus it includes the foundation of mechanics; variational formulations; computational mechanics; statics, kinematics and dynamics of rigid and elastic bodies; vibrations of solids and structures; dynamical systems and chaos; the theories of elasticity, plasticity and viscoelasticity; composite materials; rods, beams, shells and membranes; structural control and stability; soils, rocks and geomechanics; fracture; tribology; experimental mechanics; biomechanics and machine design. The median level of presentation is the first year graduate student. Some texts are monographs defining the current state of the field; others are accessible to final year undergraduates; but essentially the emphasis is on readability and clarity.

Springer and Professors Barber and Klarbring welcome book ideas from authors. Potential authors who wish to submit a book proposal should contact Dr. Mayra Castro, Senior Editor, Springer Heidelberg, Germany, email: mayra.castro@springer.com

Indexed by SCOPUS, Ei Compendex, EBSCO Discovery Service, OCLC, ProQuest Summon, Google Scholar and SpringerLink.

Uwe Mühlich

Enhanced Introduction to Finite Elements for Engineers

Springer

Uwe Mühlich
Facultad de Ciencies de Ingeniería
Universidad Austral de Chile
Valdivia, Chile

ISSN 0925-0042 ISSN 2214-7764 (electronic)
Solid Mechanics and Its Applications
ISBN 978-3-031-30424-8 ISBN 978-3-031-30422-4 (eBook)
https://doi.org/10.1007/978-3-031-30422-4

This Springer imprint is published by the registered company Springer Nature Switzerland AG
The registered company address is: Gewerbestrasse 11, 6330 Cham, Switzerland

Preface

A simple Internet search reveals that a large number of textbooks and scripts on the subject already exist. Why another one?

The Finite Element Method (FEM) is nowadays the preferred approach in many branches of science and engineering for solving numerically boundary value problems. Therefore, there is not just one sort of audience. On the contrary, there are many and they differ considerably in professional background, interest, and particular motivation to learn about the subject in the first place. In addition, there are different ways to introduce the method to students, Ph.D. students, working professionals, etc.

Furthermore, it is impossible to cover all aspects of the method in an introductory course. Hence, lecturers have to make choices according to student's background, time constraints, own preferences, etc. The selection made here reflects the author's opinion as to what knowledge is required to start applying the method safely to engineering problems. Furthermore, any introductory text should put students in the position to explore more advanced topics independently if necessary. In this context, it seems reasonable to stimulate student's interest in the mathematical underpinnings of the method.

Applying FEM safely requires not only knowledge about the method but also a sound understanding of the boundary value problem. Therefore, FEM is often discussed by means of examples from a specific area which makes it also easier to illustrate pitfalls and possible sources of errors when applying the method. This book is based in large parts on lecture notes of courses about Finite Elements taught mostly at TU-Bergakademie Freiberg, University of Antwerp, and Universidad Austral de Chile, to students in mechanical engineering, computational material science, construction technology, and construction engineering. Therefore, this is also the audience targeted here. Given the interdisciplinary character of many real-life engineering projects, however, it seems beneficial to emphasize on the universality of the method with respect to boundary value problems in science and engineering.

Limiting an introductory text on the subject to linear and time-independent problems does not seem to be appropriate anymore. There is an increasing need and interest in non-linear and time-dependent FEM solution schemes as well as in FEM for multi-physics problems already in introductory courses.

The principle objectives of the book can be defined as follows. As every introductory text, it aims to lay out the fundamentals of the method as comprehensible as possible. The book emphasizes on the universality of the method with respect to boundary value problems in science and engineering. The book aims to educate explicitly towards critical revision of results based on knowledge about boundary value problem and numerical method. It is an explicit purpose of the book to go beyond the application of the method for linear time-independent problems. Furthermore, it aims to provide background knowledge regarding underlying mathematical concepts.

I would like to express my appreciation to all students who supported one of the courses on the subject mentioned above in terms of questions, feedbacks, and critical remarks. I am also grateful to all participants of the one-week intensive courses on Finite Element Methods for non-linear boundary value problems, organised at SCKCEN Mol (Belgium) and Yildiz Technical University, Istanbul (Turkey) in 2016, for their interest and valuable contributions in the discussions. Special thanks as well to Lou Areia, Seetharam Suresh, Turgut Kocaturk, and Mesut Simsek for organising these opportunities.

Furthermore, I would like to thank Springer Scientific Publishing, in particular Mayra Castro, Saravan Mano Priya and Selvaraj Vijay Kumar, for the support provided during the process of publishing.

Valdivia, Chile
March 2023

Uwe Mühlich

Contents

Symbols

$\mathbb{N}$	Set of natural numbers
$\mathbb{Z}$	Set of integer numbers
$\mathbb{Q}$	Set of rational numbers
$\mathbb{R}$	Set of real numbers
$\mathbb{C}$	Set of complex numbers
$\bigcap$	Intersection of sets: $A \bigcap B = \{x \mid x \in A \text{ and } x \in B\}$
$\bigcup$	Union of sets: $A \bigcup B = \{x \mid x \in A \text{ or } x \in B\}$
$\subseteq$	Subset
$\subset$	Proper subset: $A \subset B$ excludes $A = B$
A^c	Complement of a set A
∂A	Boundary of set A
$\text{cl}(A)$	Closure of a set, $\text{cl}(A) = A \cup \partial A$
$\{a, b, \ldots\}$	Finite or countable set
$A \to B$	Mapping from set A to set B
Ω	Domain
$\boldsymbol{u}$	Vector
$\boldsymbol{e}_i$	i-th base vector in $\mathbb{R}^N$
T	General tensor
$\underset{\sim}{\mathbf{A}}$	Skew-symmetric tensor
$\wedge$	Exterior product
δ_{ik}	Kronecker symbol
f'	Short hand notation for $\frac{\mathrm{d}f}{\mathrm{d}x}$ or $\frac{\partial f}{\partial x}$, depending on the context
$\dot{f}$	Short hand notation for $\frac{\mathrm{d}f}{\mathrm{d}t}$ or $\frac{\partial f}{\partial t}$, depending on the context
$\frac{\partial^k}{\partial x_i^k}$	k-th partial derivative with respect to x_i
$f_{,u}$	Short hand notation for the partial derivative of f with respect to u
$\int_\Omega \Phi \,\mathrm{d}V$	Volume integral of Φ over the domain Ω
$\int_{\partial\Omega} \Phi \,\mathrm{d}A$	Surface integral of Φ over the boundary of the domain $\partial\Omega$
$\int_\Omega \Phi \,\mathrm{d}^N x$	Integral of Φ over the domain $\Omega \subset \mathbb{R}^N$ using coordinates x_i, $i = 1, .., N$
$C^k(\Omega)$	Space of all k-times continuously differentiable functions defined on the domain Ω

$L(\Omega)$	Space of all Lebesgue integrable functions defined on the domain Ω
$\sum$	Summation symbol, for instance, $\Sigma_{i=0}^{3} a_i = a_0 + a_1 + a_2 + a_3$
$\prod$	Product symbol, for instance, $\Pi_{i=0}^{3} a_i = a_0 a_1 a_2 a_3$
$\underline{\underline{A}}$	Matrix
$\underline{a}$	Matrix of one column
symm	Symmetric

Chapter 1
Introduction

Abstract After defining the Finite Element Method concisely as a method for solving boundary value problems numerically based on a weak form approach, one dimensional heat transport is used to illustrate the meanings of "boundary value problem", "initial boundary value problem" and "strong form". The structure of the book is presented and contemporary challenges regarding the education of engineers in FEM are discussed emphasising on the importance of well synchronized theory lectures, hand-calculation exercises and carefully designed computer labs.

1.1 Boundary Value Problems and Their Strong Forms

The Finite Element Method (FEM) is nowadays one of the most used analysis tools in engineering. In a nutshell, the method can be described by the following statement.

> The Finite Element Method is a tool for solving numerically boundary value problems based on a weak form approach.

Of course, this definition immediately gives rise to the following questions.

1. What is a boundary value problem?
2. What is meant by weak form approach?
3. Since there is seemingly a weak form, are there other forms as well?
4. Why FEM is based on a weak form approach?

We will use a simple example, more specifically, one dimensional heat transport, to illustrate first the meaning of the terms boundary value problem and strong form. In addition, the meaning of initial boundary value problem is explained. The remaining questions will be answered in more detail in the next chapter.

U. Mühlich, *Enhanced Introduction to Finite Elements for Engineers*, Solid Mechanics and Its Applications 268, https://doi.org/10.1007/978-3-031-30422-4_1

Imagine a wall as shown in Fig. 1.1. For the sake of simplicity, we assume, that the temperature θ does not depend on the vertical position in the wall. The horizontal position in the wall is given by the coordinate x. In general, temperature is a function of location x and time τ hence, $\theta = \theta(x, \tau)$. In a stationary state, i.e., when temperature in the wall does not depend on time, θ obeys inside the wall the differential equation

$$(\lambda(x)\theta'(x))' = 0 \tag{1.1}$$

where the prime is a short hand notation for the derivative with respect to x and λ is a material parameter, more specifically, the heat conduction coefficient of the material, the wall is made of. If λ is constant, i.e., $\lambda(x) = \lambda_0$, the solution of this equation can be obtained easily by direct integration

$$\theta'(x) = c_0 \tag{1.2}$$

$$\theta(x) = c_0 x + c_1 \tag{1.3}$$

with the integration constants c_0 and c_1. These constants have to be determined from values given at the left and right boundary of the wall, hence the name boundary value problem. If the temperatures at both sides are constant, for instance,

$$\theta(0) = 0 \tag{1.4}$$

$$\theta(l) = \theta_l\,, \tag{1.5}$$

we obtain $c_1 = 0$, $c_0 = \theta_l / l$ and, therefore

$$\theta(x) = \theta_l \frac{x}{l}\,. \tag{1.6}$$

Equations (1.1) together with (1.4) and (1.5) define the so-called strong form of the boundary value problem. Because the unknown function θ depends only on one real variable, the prime denotes an ordinary derivative and (1.1) is an ordinary differential equation. Because the highest order of derivation determines the order of a differential equation, (1.1) is a second order ordinary differential equation. Furthermore, it is a linear differential equation because the unknown function itself and its derivatives appear only linearly.

As indicated in Fig. 1.1, idealizing reality by a boundary value problem implies usually the intermediate step of idealizing real materials as continuous, i.e., replacing real matter by an abstract continuum.

Another problem with a completely different meaning, namely, an elastic rod with constant stiffness λ_0 is defined by the differential equation

$$\lambda u''(x) = 0 \tag{1.7}$$

provided, that there is no line load, and boundary conditions, for instance,

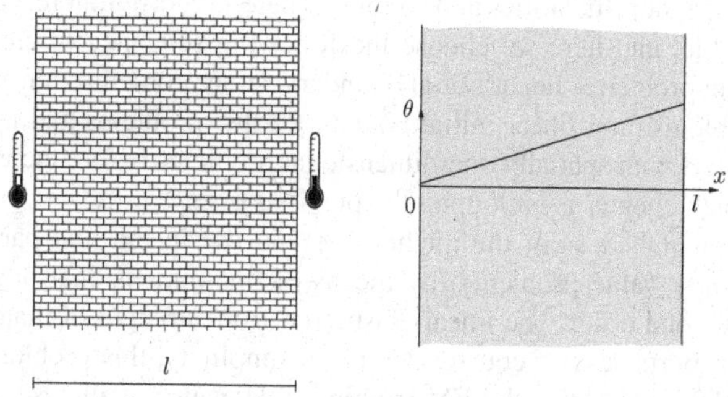

Fig. 1.1 Heat transport through a wall of thickness l. The "real" situation is sketched left, whereas the model is depicted at the right

$$u(0) = 0 \tag{1.8}$$
$$u(l) = u_l\,. \tag{1.9}$$

Here, the function to be determined is the axial displacement $u(x)$. The interesting observation is, that both problems are described essentially by the very same equations. The only difference is the specific meaning of the variables.

Let us go back to heat transport in a wall. Before reaching a stationary state, the temperature distribution undergoes an evolution in time starting from some initial state. Such a non-stationary transport problem is captured by the differential equation

$$\dot{\theta} - k\theta'' = r \tag{1.10}$$

where the dot indicates the derivative with respect to time and r is a source or sink term. The equation describes how the temperature at a given location in the wall changes with time due to heat transport and heat supply or withdrawal at this point in the presence of some source or sink, respectively. Because θ is now a function of location and time, dot and prime are actually partial derivatives and Eq. (1.10) is a partial differential equation. Its solution requires not only boundary conditions but also information about the initial state, e.g., $\theta(x, t = 0) = \theta_0$. Hence, the name initial boundary value problem.

1.2 The Structure of the Book

Needless to say, that, as in most textbooks, information is structured according to complexity starting from the simplest case and proceeding gradually towards more and more complex situations. However, there are various aspects which affect the complexity of a problem, such as, spatial dimension, order of involved derivatives, time dependence, non-linearities and so forth. Therefore, different reasonable orderings are possible, and here we choose the following sequence of chapters: linear boundary value problems, linear initial boundary value problems, non-linear boundary value problems, non-linear initial boundary value problems and multi-physics. All chapters start with spatially one-dimensional problems before increasing spatial dimension. Each chapter is built upon the previous ones.

The fundamental ideas of the method are laid out in the first part of Chap. 2 (Linear boundary value problems) by means of one-dimensional boundary value problems of second order. The linear elastic rod from strength of materials is used for illustration purposes. Of course, due to its simplicity, this problem is hardly a candidate for being solved with FEM in practice. However, it allows to focus on the fundamental principals of the method first without much risk of being distracted by technicalities unavoidable for discussing more complex boundary value problems. On the contrary, only a rather limited mathematical background, more specifically, real functions, ordinary differential equations and integration in $\mathbb{R}$, is required at this stage.

After having a sound understanding of the fundamentals, the complexity of the boundary value problem under consideration is increased. As the complexity of the boundary value problem increases, the complexity of the finite element method increases to the same extent. The Euler-Bernoulli beam, for instance, requires additional effort due to continuity requirements. Increasing spatial dimension introduces new challenges.

Chapter 3 discusses the implications of time dependence for applying FEM to linear time dependent problems. Non-linear boundary value problems are discussed in Chap. 4 and the knowledge from Chaps. 4 and 3 is combined in the first part of Chap. 5 which discusses FEM for non-linear initial boundary value problems. Eventually, the case of several coupled initial boundary value problems, i.e., multiphysics, is outlined in the second part of Chap. 5.

The reader is supposed to possess an education in strength of materials or similar and, as well, knowledge in engineering mathematics. However, since this book intends to be as self-contained as possible, elements of linear algebra and real analysis which are important in the context of FEM are provided in Appendix A and Appendix B, respectively. A condensed introduction to functional analysis is given in Appendix C. Last but not least, solutions of selected problems can be found in Appendix D.

1.3 Challenges in Learning and Teaching the Method in an Engineering Context

Engineering design in terms of functionality, safety, serviceability and durability commonly relies on models encoded in the language of mathematics. FEM, as already mentioned, is essentially a mathematical method for finding approximate solutions for models generally classified as boundary value problems.

Given the development over the last decades, FEM is nowadays rather a technology than just a method. This, however, bears the risk of generating the illusion of an almost infallible tool, which can be used without sound background knowledge and the author cannot agree more with the following statement.[1]

> "The key to good analyses is knowledge of the limitations of the method and an understanding of the physical phenomena under investigation."

A course about FEM should therefore be well-balanced between theory lectures, hand calculation exercises and computer labs but not necessarily with this rigid sequence. Coached hand calculation exercises are not only useful for illustrating theoretical background. They allow to focus on the method without being distracted by software handling issues. The sequence of steps performed manually for carefully designed exercises helps students to understand better the internal logic of FEM software and the information to be provided to FEM programs. Hand calculation exercises are very useful to stimulate students to dissect a problem first before starting the calculation in order to compute as efficiently as possible, for instance, by taking advantage of symmetries, superposition, etc. The mathematical consequences of missing essential boundary conditions are far more transparent than a maybe even rather unspecific error message from a FEM software. Of course, keeping hand calculations manageable and didactically useful requires a careful design of exercises. Combining such exercises with computer algebra systems like Maple, Mathematica or Maxima is also an option. It should be noted as well, that symbolic results are far more informative regarding, for instance, the effect of certain parameters on the result, than the floating point representation of a number.

Obviously, a serious education in FEM cannot be done without computer labs. However, getting a simulation running and visualising some results is at most thirty percent of the task! Equally important is to defend or to reject results based on thoroughly performed plausibility checks. Computer labs in FEM must be far more than just a course in software handling. It is also recommendable to use multi-purpose FEM software instead of specialized programs which already include load combinations, safety factors, etc., according to some standard procedures or safety codes. Computer labs must not be limited to show how it works but to the same extend they should demonstrate what can go wrong.

It is weird, to say the least, that there are still computer labs in which students are treated in pretty much the same way as it was done twenty years ago ignoring completely the progress in software technology and the existence of the internet. Twenty years ago, one of the main tasks of computer labs was to help people to overcome their fear of computers. Nowadays students are growing up naturally with

---

[1] The Open University, Introduction to finite element analysis.

computer technology and carry permanently a smart phone, i.e., a computer with up to date hardware and software technology. A step by step introduction to the software during which students just copy from a projection screen a sequence of steps performed by an instructor can perfectly be replaced nowadays by an individual exploration of the software using tutorials from the internet in combination with knowledge gained in previous lectures and hand calculation exercises. Of course, this requires access to the software also outside the computer lab. Students do not need nowadays much help in software handling. What is needed are carefully designed computer exercises by which students become fully aware, for instance, of the GIGO (**G**arbage **I**n—**G**arbage **O**ut) principle. A number of such exercises are provided in this book.

Chapter 2
Linear Boundary Value Problems

Abstract The chapter describes in detail the Galerkin Finite Element Method (FEM) for time independent, linear boundary value problems. Such boundary value problems can be found in many areas like linear elasticity, linear stationary transport of heat or matter, respectively, or linear electrostatics, just to mention a few. The key ideas of the method are laid out in the first part of this chapter for the Poisson equation in one dimension. Starting from the strong form of the boundary value problem, weak derivative, variational form, and weak form of the boundary value problem are introduced. The latter provides a generic way to derive FEM solution schemes, together with the idea of piecewise defined trial and test functions, which is also rigorously adapted throughout the book. The transition to two or three spatial dimensions is illustrated by means of the Poisson equation in $\mathbb{R}^N$. Necessary concepts from math are briefly revisited. Linear elasticity introduces new challenges due to the fact, that primary unknowns are no longer scalar fields but vector fields. Advanced continuity requirements are discussed for fourth order boundary value problems using the Euler-Bernoulli beam as illustrative example.

2.1 The Poisson Equation in $\mathbb{R}$

2.1.1 Strong Form

The problems of engineering and physics considered here are supposed to occur in a space with three dimensions. Hence, corresponding models are in general spatially three-dimensional. However, under certain conditions, reality can be captured with sufficient accuracy by two-dimensional or even one-dimensional models.

In the following, the elastic rod is used as an example to introduce fundamental aspects of FEM. Readers who prefer to discuss the subject in a different context such as stationary heat/mass transport, frictionless flow through pipes or electrostatics are referred to Table 2.1 for the corresponding analogies.

U. Mühlich, *Enhanced Introduction to Finite Elements for Engineers*, Solid Mechanics and Its Applications 268, https://doi.org/10.1007/978-3-031-30422-4_2

Table 2.1 Some applications covered by the Poisson equation in $\mathbb{R}$ together with the corresponding meaning of involved variables

	Axial deformation of a linear elastic rod	Linear stationary transport of heat or mass	Frictionless pipe flow	Electrostatics
u	Axial displacement	Temperature or concentration	Hydrostatic pressure	Electrostatic potential
P	Axial force	Heat or mass flux	Flow rate	Electric flux
ε	Strain	Temperature or concentration gradient	Pressure gradient	Potential gradient
λ	Stiffness	Heat or mass conductivity	Flow resistance	Dielectric constant
n	Axial line load	Heat or mass source distribution	Source distribution	Charge density

Example 2.1 We consider a rod (also called truss) of linear elastic material, length l and stiffness λ. The latter is the product of Young's modulus and cross section area. The rod is fixed at one end and exposed to a line load n. A concentrated force P_l is prescribed as shown in the figure below.

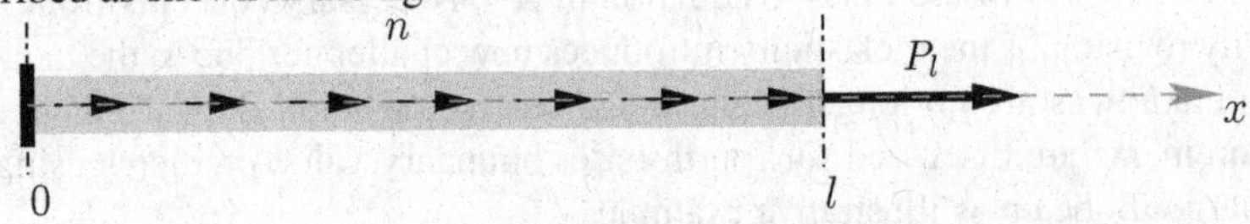

The axial force P is the product of cross section area and normal stress. The model assumes a constant stress distribution over the cross section. Axial displacement and the axial strain are labelled by u and ε, respectively. All quantities, except the axial force P_l at the right boundary, are in general functions of the position within the rod.

To convert the problem addressed in Example 2.1 into mathematical equations, a coordinate called x is introduced such that the rod occupies a domain Ω indicated by the interval $\Omega = (0, l)$. The boundary value problem is captured by the following equations, which have to be fulfilled everywhere inside the domain Ω:

- equilibrium condition: $$P' + n = 0 \tag{2.1}$$
- kinematics: $$\epsilon = u' \tag{2.2}$$
- constitutive relation (material law): $$P = \lambda\,\epsilon \tag{2.3}$$

where a prime is used to indicate the derivative with respect to the spatial coordinate x as long as it does not cause ambiguity or confusion.

Equations (2.1)–(2.3) can be condensed into a single equation. It follows from material law (2.3) and kinematics (2.2), that

$$P = \lambda u' \tag{2.4}$$

holds and plugging this result into the equilibrium condition (2.1) yields

$$(\lambda\, u')' + n = 0 \tag{2.5}$$

which is an ordinary differential equation of second order for the unknown axial displacement u. Equation (2.5) is known as Poisson equation in one dimension.

The order of this differential equation implies, that two boundary conditions are required to determine its solution. The boundary conditions for the Example 2.1 read:

$$u(0) = 0$$
$$P(l) = P_l$$

which illustrates, that different types of boundary conditions are possible.

Prescribing the value of the unknown function at the boundary is known as Dirichlet boundary condition. Imposing a value for the flux $\lambda u'$, which in the case of the rod corresponds to an axial force, is called Neumann boundary condition. Depending on whether or not prescribed values are equal to zero, boundary conditions are called homogeneous and non-homogeneous, respectively.

Equations (2.5) together with a complete set of boundary conditions is called the strong form of the boundary value problem. The task of solving the boundary value problem which corresponds to Example 2.1 can be expressed provisionally as follows.

Task 2.1 Given the functions $\lambda(x)$ and $n(x)$. Find a function $u(x)$ such that

$$\begin{aligned} (\lambda u')' + n &= 0 \qquad \text{for } x \in \Omega = (0, l)\,, \\ u(0) &= 0, \\ \lambda(l) u'(l) &= P_l\,. \end{aligned}$$

One particular drawback of the strong form is illustrated using the following example mainly for two reasons: firstly, to motivate the weak form of the boundary value problem, and secondly, to emphasize the need for a more precise language.

Example 2.2 We consider Example 2.1 with a stiffness λ given by

$$\lambda(x) = \lambda_0 \begin{cases} 1 & x \in (0, a] \\ \beta & x \in (a, l) \end{cases}$$

and $n(x) = 0$ for simplicity. The positive real number $\beta \neq 1$ is supposed to be given. According to product rule, (2.5) reads $\lambda' u' + \lambda u'' = 0$, but λ is not differentiable at $x = a$ because taking the derivative of λ from the right at $x = a$ yields

$$\frac{\mathrm{d}}{\mathrm{d}x}\lambda(x)\bigg|_{x=a_+} = \lim_{\Delta x \to 0_+} \frac{\beta - 1}{\Delta x} \to \infty\,.$$

Therefore, task (2.1) does not make sense, even though, the considered variation in material properties seems perfectly reasonable from an engineering point of view. A common way to remedy this is to modify the task accordingly. The domain Ω is split into two subdomains $\Omega_1 = (0, a), \Omega_2 = (a, l)$. The boundary value problem is solved separately for the two subdomains and the final solution is obtained by demanding continuity of u and the flux $\lambda u'$ at the interface, i.e., at $x = a$, see Exercise 2.1.

Even though, the procedure described in Example 2.2 is relatively easy and convenient for spatially one-dimensional problems, it is not hard to imagine how cumbersome this can become in two or three-dimensional situations with discontinuities along curved lines or surfaces. Therefore, the question arises if there are alternatives to (2.5) which allow to incorporate discontinuities somehow generically. Such alternatives exist and they are strongly related with the so-called weak derivative to be discussed within the next section.

2.1.2 The Weak Derivative

It seems convenient to recall, that the classical or strong derivative of some function $f(x)$

$$\frac{\mathrm{d}}{\mathrm{d}x} f(x) = f'(x) = \lim_{h \to 0} \frac{1}{\Delta x}[f(x+h) - f(x)] \tag{2.6}$$

is defined locally. The derivative f' exists in the entire domain $\Omega = (a, b)$, if for every $x \in \Omega$ a unique limit exists. The simple example $f(x) = x^2$ illustrates the procedure

$$f'(x) = \lim_{h \to 0} \frac{1}{h}[(x+h)^2 - x^2] = 2x\,.$$

But, considering for instance $f(x) = |x|, x \in \Omega = (-1, 1)$, the task of finding f' in Ω does not make sense because the classical derivative is not continuous at $x = 0$. Splitting Ω into two subintervals $\Omega_1 = (-1, 0)$ and $\Omega_2 = (0, 1)$ does not solve the problem because, even though $f' = -1$ for $x \in \Omega_1$ and $f' = +1$ for $x \in \Omega_1$, it is not clear which value to assign at $x = 0$.

Therefore, the question arises, if the notion of differentiability can be generalized for a wider range of functions, such that it contains the strong derivative as a special case. The key for solving this problem is the well-known integration by parts theorem. If the classical derivative of a function f exists in the domain $\Omega = (a, b)$ then

$$\int_a^b f' v \, \mathrm{d}x = -\int_a^b f v' \, \mathrm{d}x + f(b)v(b) - f(a)v(a)$$

holds for any continuously differentiable function v. The boundary terms can be omitted if v vanishes at the boundary points, i.e., $v(a) = v(b) = 0$. With this restriction in terms of v, the integration by parts theorem reads

$$\int_a^b f' v \, \mathrm{d}x = -\int_a^b f v' \, \mathrm{d}x \,. \tag{2.7}$$

As long as f' exists in Ω and v vanishes at the boundary points, (2.7) is just an identity. However, considering the right hand side of (2.7) as meaningful on its own right, the weak derivative of a function f can be defined as follows. A function g which fulfils

$$\int_a^b g v \, \mathrm{d}x = -\int_a^b f v' \, \mathrm{d}x \tag{2.8}$$

for arbitrary but continuously differentiable test functions v with $v(a) = v(b) = 0$, is called the weak derivative of f. By abuse of notation, $g = f'$ is commonly used if the meaning of the derivative is clear from the context. Otherwise, an explicit notation, e.g., $g := Df$, is employed. The concept of the weak derivative extends easily to higher derivatives.

Example 2.3 (*Weak derivative of $f(x) = x^2$*) According to (2.8) we have

$$\int_a^b g v \, \mathrm{d}x = -\int_a^b x^2 v' \, \mathrm{d}x \,.$$

Applying integration by parts for the right hand side yields

$$\int_a^b g v \, \mathrm{d}x = -\left[x^2 v\Big|_a^b - \int_a^b 2x \, v \, \mathrm{d}x \right] .$$

Since, the boundary terms vanish due to the definition of v, $g = 2x$ is readily obtained by comparing the arguments of the integral operators. Because $f(x) = x^2 \in C^1(\Omega)$, classical and weak derivative coincide.

Example 2.4 (*Weak derivative of $f(x) = |x|$*) As discussed above, there is no classical derivative of $f(x) = |x|$ in the interval $(-1, 1)$ because the limit in (2.6) is not unique for $x = 0$. However, the equation

$$\int_{-1}^{1} g v \, \mathrm{d}x = -\left[\int_{-1}^{0} (-x)v' \, \mathrm{d}x + \int_{0}^{1} (+x)v' \, \mathrm{d}x\right] = \left[\int_{-1}^{0} -v \, \mathrm{d}x + \int_{0}^{1} v \, \mathrm{d}x\right]$$

is fulfilled by

$$g = \begin{cases} -1 & x \in (-1, 0) \\ \beta \quad \text{if} & x = 0 \\ +1 & x \in (0, 1) \end{cases} \tag{2.9}$$

where β is some arbitrary real number, i.e., $\beta \in \mathbb{R}$. Therefore, (2.9) is the weak derivative of the function $f(x) = |x|$ in the interval $(-1, 1)$. The appearance of β clearly requires an explanation. It shows, that the weak derivative is far more profound than it might seem at first glance. For more detailed information, see Sect. B.8 of Appendix B.

The function g in (2.9) equals the Heaviside-function $H(x - a)$ for $a = 0$ in $\Omega = (-1, 1)$. The general definition of $H(x - a)$ reads

$$H(x - a) = \begin{cases} -1 & x < a \\ \beta & \text{if } x = a \\ +1 & x > a \end{cases} \tag{2.10}$$

with arbitrary constant β.

2.1.3 *Variational and Weak Form*

Inspired by the idea behind the weak derivative, (2.5) is multiplied by an arbitrary but sufficiently continuous test function v and the result is integrated over the considered domain, which yields

$$\int_{0}^{l} (\lambda u')' v \mathrm{d}\, x + \int_{0}^{l} n \, v \mathrm{d}\, x = 0 \,. \tag{2.11}$$

Applying integration by parts to the first term of the left hand side of (2.11) gives

$$\int_0^l \lambda u'v'\mathrm{d}x + (\lambda u)'(0)v(0) - (\lambda u)'(l)v(l) - \int_0^l n\, v\mathrm{d}x = 0\,. \tag{2.12}$$

Taking into account (2.4), it can readily be seen that the third term on the left hand side of (2.12) is directly related to the Neumann boundary condition $\lambda(l)u'(l) = P_l$. If the test function only vanishes at boundaries with Dirichlet boundary conditions, the Neumann boundary condition can be plugged into (2.12), which yields the following system of equation

$$\int_0^l \lambda u'v'\,\mathrm{d}x - \int_0^l n\,v\,\mathrm{d}x - P_l v(l) = 0 \tag{2.13}$$

$$u(0) = 0 \tag{2.14}$$

known as variational form of the boundary value problem (2.1).

For twice continuously differentiable functions, the variational form given by (2.13) and (2.14) is equivalent to the strong form defined by (2.1). This can readily be shown by applying integration by parts just backwards and is left as an exercise to the reader.

Since Neumann boundary conditions can be incorporated directly into the variational form of (2.5) they are also called "natural" boundary conditions. Dirichlet boundary conditions, on the other hand, are known as "essential" boundary conditions.

The results obtained so far suggest a general methodology for deriving the variational form. It consists of the following steps which are illustrated by means of Example 2.1:

1. multiplication of the differential equation of the strong form with a test function and integration with respect to the considered domain

$$\int_0^l [(\lambda u')' - n]v\,\mathrm{d}x = 0$$

2. integration by parts taking into account, that test functions vanish at boundaries with prescribed essential boundary conditions, e.g.,

$$\int_0^l \lambda u'v'\,\mathrm{d}x - \int_0^l n\,v\,\mathrm{d}x - \lambda(l)u'(l)v(l) = 0$$

3. incorporation of natural boundary conditions in the result of the previous step and finalizing the procedure by adding essential boundary conditions

$$\int_0^l \lambda u'v'\,\mathrm{d}x - \int_0^l n\,v\,\mathrm{d}x - P_l v(l) = 0$$

$$u(0) = 0$$

Remark 2.1 (*Principle of virtual displacements*) Denoting displacements and virtual displacements, as usual, by u and δu, respectively and substituting v by δu in the weak form yields the principle of virtual displacement. The latter is often used in engineering mechanics as starting point for FEM.

Example 2.5 To demonstrate the use of the variational form for solving boundary value problems, a rod of length l with homogeneous Dirichlet boundary conditions exposed to a constant line load n_0 is considered. For this case (2.12) reads

$$\frac{\lambda_0}{n_0}\int_0^l u'\,v\,\mathrm{d}x = \int_0^l 1\,v\,\mathrm{d}x .$$

To be able to compare the arguments of the integrals of both sides, we recast the term of the right hand side properly and apply integration by parts

$$\frac{\lambda_0}{n_0}\int_0^l u'\,v\,\mathrm{d}x = \int_0^l (x + d_1)'\,v\,\mathrm{d}x$$

$$= [(x + d_1)v]_0^l - \int_0^l (x + d_1)\,v'\,\mathrm{d}x .$$

The boundary terms vanish due to the properties of the test function. Because of the arbitrariness of the test function, the equation above can only be fulfilled if

$$u'(x) = -\frac{n_0}{\lambda_0}(x + d_1)$$

where d_1 is an arbitrary constant. Direct integration taking into account the boundary conditions finally gives

$$u(x) = \frac{n_0 l^2}{2\lambda_0}\left[\frac{x}{l} - \left(\frac{x}{l}\right)^2\right]$$

and this solution coincides with the result obtained by direct integration of the corresponding strong form which is left as an exercise to the reader.

So far, the variational form of the boundary value problem has been introduced just as an equivalent alternative to the strong form. A significant conceptual leap is achieved by the following idea.

> The variational form can be considered as meaningful on its own right, regardless of the existence of a strong form.

This opens a completely new view on the subject and allows for milder restrictions regarding continuity as illustrated by the following example.

Example 2.6 To illustrate the use of the variational form in the presence of discontinuities in the source term, we consider a domain $\Omega = (0, l)$ together with a source term

$$n(x) = n_0 \begin{cases} 0 & x \in (0, a) \\ 1 & x \in [a, l) \end{cases}$$

with $\lambda(x) = \lambda_0$, $u(0) = 0$ and $\lambda_0 u'(l) = 0$. Applying the same methodology used for Example 2.5, the variational form can be written as

$$\int_0^l u'v' \, \mathrm{d}x = \frac{n_0}{\lambda_0} \left\{ \int_0^a d_1' v \, \mathrm{d}x + \int_a^l [x + d_2]' v \, \mathrm{d}x \right\} .$$

Integration by parts taking into account, that the test function vanishes if essential boundary conditions are prescribed, yields

$$\begin{aligned} \int_0^l u'v' \, \mathrm{d}x = \; & \frac{n_0}{\lambda_0} \{ d_1 v(a) + [l + d_2] v(l) - [a + d_2] v(a) \} \\ & - \frac{n_0}{\lambda_0} \left\{ \int_0^a d_1 v' \, \mathrm{d}x + \int_a^l [x + d_2] v' \, \mathrm{d}x \right\} . \end{aligned}$$

The boundary terms must vanish, which implies $d_2 = -l$ and $d_1 = a - l$. Therefore,

$$u' = \frac{n_0 l}{\lambda_0} \begin{cases} 1 - \frac{a}{l} & x \in (0, a) \\ 1 - \frac{x}{l} & x \in (a, l) \end{cases}$$

is obtained, which shows, that u is not twice continuously differentiable. Direct integration, taking into account the homogeneous Dirichlet boundary condition at $x = 0$ and demanding continuity of u at $x = a$ gives eventually

$$u = \frac{n_0 l^2}{\lambda_0} \begin{cases} [1 - \frac{a}{l}]\frac{x}{l} & x \in (0, a) \\ \frac{x}{l} - \frac{1}{2}\left[\left(\frac{x}{l}\right)^2 + \left(\frac{a}{l}\right)^2\right] & x \in [a, l) \end{cases} .$$

Remark 2.2 As illustrated by Examples 2.2 and 2.6, discontinuities in the source term require to divide the interval into subintervals regardless which form is used. Therefore, the reader may question the advantage of using the variational form rather than the strong form. However, a detailed comparison of the solution steps reveals, that the type of compatibility condition required in addition to the boundary conditions for determining the final solution is different in both cases. This comparison as well as the corresponding conclusions are left as an exercise to the reader.

Even more can be done. The variational form still depends on the existence of the classical first derivatives and the question arises whether or not continuity requirements can be relaxed even more by interpreting these derivatives as weak derivatives. The answer is yes, which implies the following definition of the weak form.

> The weak form of a boundary value problem is derived from its variational form by interpreting derivatives as weak derivatives.

Remark 2.3 (*Weak form*) Not only can the weak form be seen as meaningful on its own right but strong and variational form of a boundary value problem are eventually special cases of the weak form under corresponding regularity conditions.

The so-called Dirac-delta distribution is rather useful for illustrating certain aspects regarding the weak form. It is defined by

$$\int_a^b \delta(x - c) f(x)\, \mathrm{d}x = f(c) \tag{2.15}$$

for $c \in (a, b)$ and maps a function $f(x)$ defined on $\Omega = (a, b)$ to its values at $x = c$. Since $\delta(x - c)$ can not be evaluated point-wise it is not seen as a function in the usual sense. It is easy to demonstrate, that

$$\int_a^b \delta(x - c) v(x)\, \mathrm{d}x = -\int_a^b H(x - c) v'(x)\, \mathrm{d}x$$

for sufficiently continuous test functions $v(x)$ with $v(a) = v(b) = 0$ which leads to the concept of a distributional derivative, i.e., $\delta(x - c) := H'(x - c)$ by abuse of notation.

Example 2.7 Homogeneous Dirichlet boundary conditions are assumed together with $\lambda(x) = \lambda_0$. As an example for a solution with only weak first derivative,

$$\int_0^l u'v'\,\mathrm{d}x = \frac{1}{\lambda_0}\int_0^l nv\,\mathrm{d}x = -\frac{n_0}{\lambda_0}\int_0^l [H(x-c)+d_1]v'\,\mathrm{d}x \qquad (2.16)$$

is considered with $c \in (0, l)$ and the Heaviside function $H(x - c)$ as defined by (2.10). The solution for $u(x)$ can readily be obtained by comparing the left hand side of (2.16) with its most right hand side and proceeding from there in a similar manner as in previous examples.

However, the relation between $H(x - a)$ and the source term $n(x)$ can only be explained using the notion of distributional derivative. The solution corresponds to the case of a point source with magnitude n_0 at $x = c$.

Remark 2.4 (*Operator notation*) Defining

$$a(u, v) = \int_\Omega \lambda u'v'\,\mathrm{d}x \qquad \text{and} \qquad b(v) = \int_\Omega n\,v\,\mathrm{d}x + P_l v(l)\,,$$

equation (2.13) can be written more concisely as

$$a(u, v) - b(v) = 0\,,$$

which is far more than a notational convenience but an important step towards a systematic and unifying treatment of boundary value problems.

2.1.4 A Precursor of FEM

In this section, a simple numerical solution scheme is developed for the variational form already given by (2.13) and (2.14)

$$\int_0^l [\lambda\, u'v' - n\, v]\,\mathrm{d}x \quad - P_l\, v(l) = 0,$$

$$u(0) = 0\,.$$

First of all, we approximate the solution $u(x)$ by a trial function $\tilde{u}(x)$ choosing a second order polynomial

$$\tilde{u}(x) = c_1 x + c_2 x^2\,, \qquad (2.17)$$

which fulfils the essential boundary condition, i.e., $\tilde{u}(0) = 0$. In addition, a representation for the test function v is required. One option is to use as well a second order polynomial, i.e.,

$$v(x) = d_1 x + d_2 x^2 . \tag{2.18}$$

Choosing a particular representation for v reduces, of course, the arbitrariness of the test function to the arbitrariness of its coefficients d_1 and d_2. Introducing $\tilde{u}(x)$, $v(x)$ into the weak form yields

$$\int_0^l \lambda[c_1 + 2c_2 x][d_1 + 2d_2 x]\,\mathrm{d}x - \int_0^l n[d_1 x + d_2 x^2]\,\mathrm{d}x - P_l[d_1 l + d_2 l^2] = 0 \tag{2.19}$$

which can be written using matrix notation as

$$[d_1\ d_2]\left\{\begin{bmatrix} 1\int_0^l \lambda\,\mathrm{d}x & 2\int_0^l \lambda\,x\,\mathrm{d}x \\ 2\int_0^l \lambda\,x\,\mathrm{d}x & 4\int_0^l \lambda\,x^2\,\mathrm{d}x \end{bmatrix}\begin{bmatrix} c_1 \\ c_2 \end{bmatrix} - \begin{bmatrix} \int_0^l n\,x\,\mathrm{d}x \\ \int_0^l n\,x^2\,\mathrm{d}x \end{bmatrix} - \begin{bmatrix} P_l l \\ P_l l^2 \end{bmatrix}\right\} = 0$$

or more compact (see, Appendix A)

$$\underline{\hat{d}}^{\mathrm{T}}\left[\underline{\underline{K}}\,\underline{\hat{c}} - \underline{q} - \underline{\hat{F}}\right] = 0 . \tag{2.20}$$

Because of the arbitrariness of $\underline{\hat{d}}$, (2.20) can only be fulfilled, if the term within the brackets vanishes. Therefore, the unknown coefficients c_1 and c_2 can be determined by solving the system of linear equations

$$\underline{\underline{K}}\,\underline{\hat{c}} = \underline{q} + \underline{\hat{F}} . \tag{2.21}$$

The scheme extends in a straight forward manner to polynomials of degree N, where the elements of $\underline{\underline{K}}$, $\underline{q}$ and $\underline{\hat{F}}$ are given by

$$K_{rs} = r\,s \int_0^l \lambda\,x^{r-1} x^{s-1}\,\mathrm{d}x , \qquad q_r = \int_0^l n\,x^r\,\mathrm{d}x , \qquad \hat{F}_r = P_l l^r , \tag{2.22}$$

with $r, s = 1, \ldots, N$. Unfortunately, as the degree of the polynomials increase, the numerical solution tends to oscillate.

Example 2.8 The variational form given by (2.13) and (2.14) is considered with $\lambda(x) = \lambda_0$, $n(x) = 0.1 n_0 \sin\left(\frac{\pi x}{l}\right)$ and $P_l = 0$.

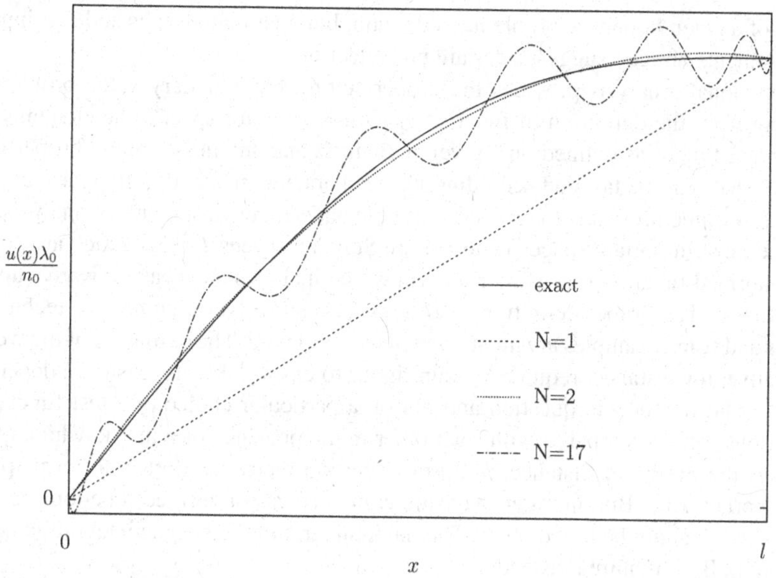

Fig. 2.1 Polynomial approximation of the variational form discussed in Example 2.8 for different polynomial degrees N together with the exact solution

Polynomial approximations of degree N are used to solve the variational form numerically according to (2.21) and (2.22). For small N it seems as if the approximation becomes better with increasing N. But, as N reaches a critical value, oscillations can be observed as shown in Fig. 2.1 which renders this approximation scheme unreliable.

In order to obtain a more robust method, a smarter choice regarding trial function and test function is required which is exactly what is done by FEM. But, as Example 2.8 illustrates, a certain rigour is required to design reliable schemes. Such rigour is provided by a discipline called Functional Analysis.

2.1.5 Comments on Functional Analysis Background

So far, essential concepts have been introduced rather informally skipping over a number of technical details. Regardless of the specific context of a given problem, a central question is always whether or not solutions even exist, and if so, how many. In the best case scenario, the problem is well-posed meaning that there is only one unique solution to the problem which depends continuously on the data in terms of material properties, source terms and boundary conditions.

Solutions of boundary value problems are functions. There are infinitely many and not every function qualifies as possible candidate for a solution. The characteristics of a given boundary problem, its domain, boundary conditions and continuity requirements already suggest a certain pre-selection.

Functional Analysis provides the proper setting for boundary value problems, starting from the definition of function spaces, i.e., vector spaces, the elements of which are functions defined on a given domain Ω and sharing common properties. Such spaces can be tailored according to the characteristics of a particular class of boundary value problems to provide a reliable and safe environment for analysis.

The most important spaces for FEM are Sobolev spaces $H^k(\Omega)$ which are complete normed function spaces with a norm which includes the weak derivative up to the order k. The proper definition of $H^k(\Omega)$ spaces involves a number of technical issues and some examples are mentioned in the following. The definition of the weak derivative, for instance, requires certain rigour to ensure, that its existence depends only on the function in question and not on a particular choice of a test function. Functions can be compared with each other in the presence of a metric which generalizes the notion of distance. A norm induces a metric and integral norms play an important role. But, in order to ensure completeness of normed function spaces, such norms cannot be based on the Riemann integral but Lebesgue integration (see, Appendix B) is required instead.

Functional Analysis is a broad field but only a limited part of it is required in the context of FEM. A very important piece of information is the so-called Lax-Milgram lemma about existence and uniquenes of solutions. Appendix C provides a minimum amount of information necessary to understand the logic behind Functional Analysis with emphasis on boundary value problems.

For the boundary value problems considered here, Functional Analysis provides a systematic and unifying treatment, which can be summarized as follows.

1. In view of the Lax-Milgram Lemma, a weak form which can be written as

$$a(u, v) = b(v)$$

   has to be derived from the strong form of the boundary value problem with trial and test functions u and v, respectively. See, as well Remark 2.4.
2. A function space V has to be defined such that $a : V \times V \to \mathbb{R}$ and $b : V \to \mathbb{R}$, taking into account as well the essential boundary conditions of the problem.
3. The question whether or not $a(u, v) = b(v)$ has a unique solution $u \in V$ for all possible $v \in V$ can now be discussed systematically based on the Lax-Milgram Lemma by analysing $a(u, v)$ and $b(v)$.

The evolution of FEM from a numerical method to a technology is not least due to the progress in Functional Analysis which nowadays provides the suitable language for FEM not only in applied mathematics but as well in engineering. Therefore, the reader is encouraged to examine in detail the information provided in Appendix C.

2.1.6 Galerkin FEM with a Piecewise Linear Global Basis

First, the problem of Example 2.1 is stated in the contemporary language of functional analysis starting with the definition of a function space $V(\Omega)$ with $\Omega = (0, l)$ in which the solution can be found,

$$V(\Omega) = \{f \mid f \in H^1(\Omega) \text{ and } f(0) = 0\}. \tag{2.23}$$

According to this definition, we are looking for a solution in the space of at least once weakly differentiable functions $f(x)$ defined over Ω and vanishing at $x = 0$. Then, the task implied by Example 2.1 reads as follows.

Task 2.2 Find $u \in V(\Omega)$ according to (2.23) such that

$$\mathcal{W} = \int_{\Omega} \lambda u' v' \, \mathrm{d}x - \int_{\Omega} n v \, \mathrm{d}x - P_l v(l) = 0$$

for arbitrary $v \in V$ and given $n, \lambda \in L^2(\Omega)$.

Remark 2.5 The concise formulation of Task 2.2 condenses all significant information. The essential boundary conditions are taken into account already by the definition of the space V. Looking for a solution in the Sobolev space $H^1(\Omega)$ implies interpreting derivatives in the weak sense and integrals as Lebesgue integrals. The latter implies that n and λ must be Lebesgue integrable as well, and, therefore in $L^2(\Omega)$.

The space $V(\Omega)$ is infinite dimensional. Galerkin FEM is based on the idea to look for a solution in a finite dimensional space $V_h(\Omega)$ which is a subspace of $V(\Omega)$, i.e., $V_h(\Omega) \subset V(\Omega)$. Task 2.2 changes accordingly as follows.

Task 2.3 Find $u_h \in V_h(\Omega)$ according to (2.23) such that

$$\mathcal{W}_h = \int_{\Omega} \lambda u_h' v_h' \, \mathrm{d}x - \int_{\Omega} n v_h \, \mathrm{d}x - P_l v_h(l) = 0$$

for arbitrary $v_h \in V_h$ and given $n, \lambda \in L^2(\Omega)$.

In the following, a particular realization of V_h is considered. First, the domain Ω is partitioned into a number of sub-domains Ω_e. The result is called a mesh. For spatially one-dimensional problems, this is equivalent with dividing the interval

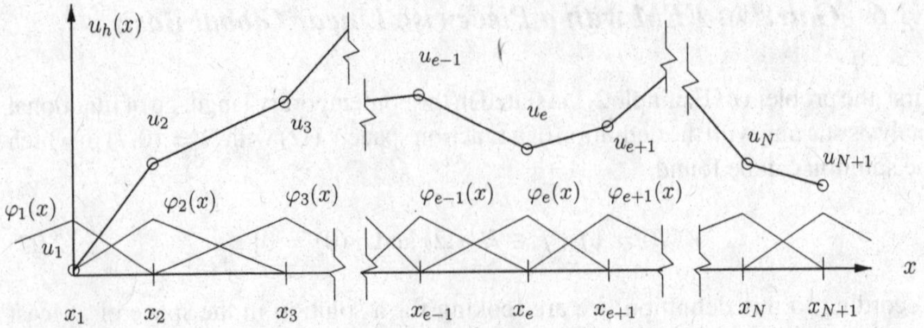

Fig. 2.2 Interpolation of a function $u_h(x)$ with known or supposed values at points x_e, by a piecewise linear function using a global basis $\{\varphi_e(x)\}$ with $e = 1, \ldots, N+1$

$\Omega = (0, l)$ into subintervalls as shown in Fig. 2.2. Every sub-domain is defined by means of nodes with coordinates x_e, where $x_{e-1} < x_e < x_{e+1}$. In order to find u_h of Task 2.3, an interpolation approach based on values of u_h at the nodes is used choosing an interpolation basis $\varphi_e(x)$ such that

$$\varphi_e(x = x_p) = \delta_{ep}$$

holds. For the meaning of the Kronecker-symbol δ_{ep}, see Definition A.17.

For the particular case of a linear interpolation approach, the corresponding basis functions are given by

$$\varphi_e(x) = \begin{cases} \frac{x - x_{e-1}}{L_{e-1}} = 1 - \frac{x_e - x}{L_{e-1}} & x \in (x_{e-1}, x_e] \\ -\frac{x - x_{e+1}}{L_e} = 1 - \frac{x - x_e}{L_e} & x \in (x_e, x_{e+1}] \\ 0 & \text{else} \end{cases}$$

with $e = 1, 2, \ldots, N+1$. Differentiation of $\varphi_e(x)$ yields

$$\varphi_e'(x) = \begin{cases} +\frac{1}{L_{e-1}} & x \in (x_{e-1}, x_e] \\ -\frac{1}{L_e} & x \in (x_e, x_{e+1}) \\ 0 & \text{else} \end{cases} .$$

A piecewise linear trial function can now be defined by the sum of basis functions multiplied by the values u_e, i.e., the values of the function at the nodes with coordinates x_e,

$$u_h(x) = \sum_{e=1}^{N+1} \varphi_e(x)\, u_e .$$

The derivative of this trial function is simply

$$u'_h(x) = \sum_{e=1}^{N+1} \varphi'_e(x)\, u_e\,.$$

The representation of the test function is constructed analogously. A more compact representation can be achieved by employing matrix notation. Therefore, we define

$$\underline{\varphi}^{\mathrm{T}} = \left[\varphi_1(x)\ \varphi_2(x)\ \ldots\ \varphi_N(x)\right]$$
$$\underline{\varphi}'^{\mathrm{T}} = \left[\varphi'_1(x)\ \varphi'_2(x)\ \ldots\ \varphi'_N(x)\right],$$

and, in addition,

$$\underline{\hat{u}}^{\mathrm{T}} = \left[u_1\ u_2\ \ldots\ u_{N+1}\right], \quad \underline{\hat{v}}^{\mathrm{T}} = \left[v_1\ v_2\ \ldots\ v_{N+1}\right].$$

A hat indicates, that the object does not contain functions but only real coefficients. This convention is used throughout the entire book.

Using the compact notation, trial function, test function and their derivatives are given by

$$u_h(x) = \underline{\varphi}^{\mathrm{T}}\,\underline{\hat{u}} \qquad v_h(x) = \underline{\varphi}^{\mathrm{T}}\,\underline{\hat{v}} \tag{2.24}$$
$$u'_h(x) = \underline{\varphi}'^{\mathrm{T}}\,\underline{\hat{u}} \qquad v'_h(x) = \underline{\varphi}'^{\mathrm{T}}\,\underline{\hat{v}} \tag{2.25}$$

and readers facing difficulties at this point are encouraged to solve corresponding exercises related to matrix operations before proceeding further.

Introducing test function and trial function into the weak form of Task 2.2 yields

$$\mathcal{W}_h = \int_{x_1}^{x_{N+1}} \lambda(\underline{\hat{v}}^{\mathrm{T}}\,\underline{\varphi}')(\underline{\varphi}'^{\mathrm{T}}\,\underline{u})\mathrm{d}\,x - \int_{x_1}^{x_{N+1}} n\,\underline{\hat{v}}^{\mathrm{T}}\,\underline{\varphi}\,\mathrm{d}\,x - \underline{\hat{F}}^{\mathrm{T}}\,\underline{\hat{v}}\,. \tag{2.26}$$

The definition of $\underline{\hat{F}}$ requires some care. $\underline{\hat{F}}$ does not just encode the term $P_l v(l)$ of Task 2.2, but

$$\underline{\hat{F}}^{\mathrm{T}} = \left[P_0\ 0\ \ldots\ 0\ P_l\right],$$

because the Dirichlet boundary condition at $x = 0$ is not yet incorporated. Therefore, at this point, all boundary terms have to be taken into account, see, as well, (2.12). However, it turns out, that P_0 drops out eventually, as shown below.

$\mathcal{W}_h$ can be further processed by taking into account, that $\underline{a}^{\mathrm{T}}\,\underline{b} = \underline{b}^{\mathrm{T}}\,\underline{a}$. Since, all hatted quantities do not contain functions but real coefficients, they can be factored out of the integrals. Hence,

$$\mathcal{W}_h = \underline{\hat{v}}^{\mathrm{T}} \left[\int\limits_{x_1}^{x_{N+1}} \lambda \underline{\varphi}'^{\mathrm{T}}\, \underline{\varphi}' \mathrm{d}\, x\, \underline{\hat{u}} - \int\limits_{x_1}^{x_{N+1}} \underline{\varphi}\, n \mathrm{d}\, x - \underline{\hat{F}} \right] = 0 \tag{2.27}$$

which can be written more concisely as

$$\mathcal{W}_h = \underline{\hat{v}}^{\mathrm{T}} \left[\underline{\underline{K}}\, \underline{\hat{u}} - \underline{q} - \underline{\hat{F}} \right] = 0\,. \tag{2.28}$$

For historical reasons, $\underline{\underline{K}}$ is called the stiffness matrix of the system, $\underline{q}$ is known as load vector and $\underline{\hat{F}}$ is the so-called reaction force vector.

The individual elements of stiffness matrix and load vector are given by

$$K_{ij} = \int\limits_{x_1}^{x_{N+1}} \lambda(x) \varphi_i'(x) \varphi_j'(x) \mathrm{d}x \tag{2.29}$$

$$q_i = \int\limits_{x_1}^{x_{N+1}} n(x) \varphi_i(x) \mathrm{d}x \tag{2.30}$$

Remark 2.6 Operations of linear algebra can be encoded by index notation or using symbolic notation. The results of an operation performed by means of symbolic notation is either already known or it has to be derived first using index notation. Afterwards it can be transferred to symbolic notation, see as well Appendix A.

Equation (2.28) demands, that the product between a transposed vector and a vector has to vanish. Because $\underline{\hat{v}}$ is arbitrary, (2.28) can only be fulfilled, if

$$\underline{\underline{K}}\, \underline{\hat{u}} = \underline{p} + \underline{\hat{F}} \tag{2.31}$$

holds, which is a system of linear equations for the unknown vector $\underline{\hat{u}}$. The stiffness matrix is singular, i.e., its determinate vanishes. Incorporating essential boundary conditions yields the constrained system of equations with non-singular stiffness matrix. Furthermore, the stiffness matrix is symmetric.

In the following, Example 2.1 is considered using a discretization of three intervals (finite elements) and $\lambda(x) = \lambda_0$. The stiffness matrix reads for this case

$$\underline{\underline{K}} = \lambda_0 \int_{x_1}^{x_{N+1}} \begin{bmatrix} \varphi_1'(x)\varphi_1'(x) & \varphi_1'(x)\varphi_2'(x) & \varphi_1'(x)\varphi_3'(x) \\ & \varphi_2'(x)\varphi_2'(x) & \varphi_2'(x)\varphi_3'(x) \\ \text{symm.} & & \varphi_3'(x)\varphi_3'(x) \end{bmatrix} \mathrm{d}x$$

and the load vector for $n(x) = n_0$ is formally given by

$$\underline{q} = n_0 \int_{x_1}^{x_{N+1}} \left[\varphi_1(x)\ \varphi_2(x)\ \varphi_3(x)\right] \mathrm{d}x\,.$$

Computation of the elements of $\underline{\underline{K}}$ is shown exemplary for K_{11}, K_{12}, and K_{22},

$$K_{11} = \int_{x_1}^{x_2} \frac{1}{L_1^2}\mathrm{d}x + 0 = \frac{1}{L_1}\,, \qquad K_{12} = \int_{x_1}^{x_2} \frac{1}{L_1}(-1)\frac{1}{L_1}\mathrm{d}x + 0 = -\frac{1}{L_1}\,,$$

$$K_{22} = \int_{x_1}^{x_2} \frac{1}{L_1^2}\mathrm{d}x + \int_{x_1}^{x_2} \frac{1}{L_1^2}\mathrm{d}x = \frac{1}{L_1} + \frac{1}{L_2}\,,$$

and the reader is encouraged to perform the remaining computations. Analogously, the elements of $\underline{q}$ evaluate to

$$q_1 = n_0 \int_{x_1}^{x_2} \varphi_1(x)\mathrm{d}x = n_0 \int_{x_1}^{x_2} \left[1 - \frac{x - x_1}{L_1}\right] = n_0 \frac{L_1}{2}$$

$$q_2 = n_0 \int_{x_1}^{x_2} \varphi_2(x)\mathrm{d}x = n_0 \int_{x_1}^{x_2} \left[1 - \frac{x_2 - x}{L_1}\right] + n_0 \int_{x_2}^{x_3} \left[1 - \frac{x - x_2}{L_2}\right] = n_0 \frac{L_1 + L_2}{2}$$

$$q_3 = n_0 \int_{x_1}^{x_2} \varphi_3(x)\mathrm{d}x + n_0 \int_{x_1}^{x_2} \left[1 - \frac{x_3 - x}{L_2}\right] = n_0 \frac{L_2}{2}$$

and the following system of linear equations is obtained

$$\lambda_0 \begin{bmatrix} \frac{1}{L_1} & -\frac{1}{L_1} & 0 \\ & \frac{1}{L_1} + \frac{1}{L_2} & -\frac{1}{L_2} \\ \text{symm.} & & \frac{1}{L_2} \end{bmatrix} \begin{bmatrix} u_1 \\ u_2 \\ u_3 \end{bmatrix} = \frac{n_0}{2} \begin{bmatrix} L_1 \\ L_1 + L_2 \\ L_2 \end{bmatrix} + \begin{bmatrix} P_0 \\ 0 \\ P_l \end{bmatrix}.$$

The essential boundary condition implies $u_1 = 0$. Therefore, the first column can be deleted. In addition the first row can be deleted too because of $v_1 = 0$, see (2.28). Eventually, the following constrained system of equations has to be solved

$$\lambda_0 \begin{bmatrix} \frac{1}{L_1} + \frac{1}{L_2} & -\frac{1}{L_2} \\ \text{symm.} & \frac{1}{L_2} \end{bmatrix} \begin{bmatrix} u_2 \\ u_3 \end{bmatrix} = \frac{n_0}{2} \begin{bmatrix} L_1 + L_2 \\ L_2 \end{bmatrix} + \begin{bmatrix} 0 \\ P_l \end{bmatrix}$$

which gives for the special case $L_1 = L_2 = l$, $P_l = 0$,

$$u_2 = \frac{3}{2}\frac{n_0 l^2}{\lambda_0}, \quad u_3 = 2\frac{n_0 l^2}{\lambda_0}.$$

Remark 2.7 Applying (2.29) and (2.30) strictly means, especially for larger systems, wasting computation time for computing explicitely a large number of zeros in the course of integration. This is usually avoided by proper case differentiation.

For spatially two-dimensional or three-dimensional problems and increasing interpolation order, the proper definition of global basis functions becomes more cumbersome.Therefore, an alternative approach is discussed in the following section.

2.1.7 *Galerkin FEM Using a Linear Local Basis*

Dividing the domain $\Omega = (0, l)$ into N sub-domains, the weak from $\mathcal{W}_h$ of Task 2.3 can be written as

$$\mathcal{W}_h = \sum_{e=1}^{N} \mathcal{W}_e - P_l v_h(l) = 0 \tag{2.32}$$

with

$$\mathcal{W}_e = \int_{x_e}^{x_{e+1}} \frac{1}{2} \lambda\, u_h' v_h' \mathrm{d}x - \int_{x_e}^{x_{e+1}} n\, v_h \mathrm{d}x\,. \tag{2.33}$$

For historical reasons, a subinterval is also called finite element. Every finite element corresponds to a domain Ω_e defined by the interval (x_e, x_{e+1}). The aim in the following is to develop a somewhat generic approach for performing as much steps as possible for one reference domain $\Omega_\square$.

Different representations of $\Omega_\square$ are possible. The most common is to consider an interval $(-1, 1)$ together with a coordinate ξ. A domain Ω_e can be related to $\Omega_\square$ by means of a coordinate transform. Since, for the case considered here, only two

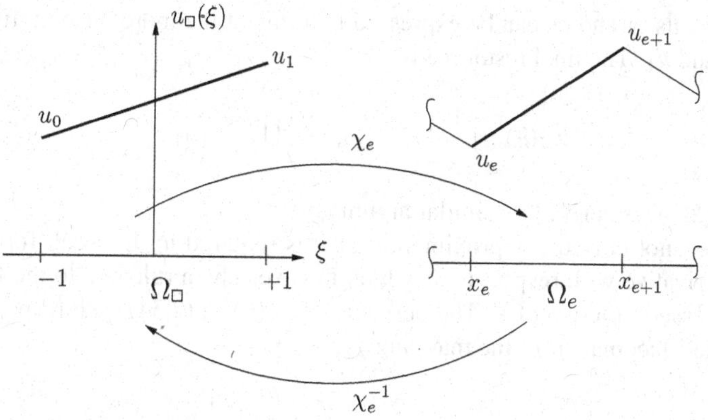

Fig. 2.3 Illustration of the reference domain $\Omega_\square$

conditions can be crafted namely $x(\xi = -1) = x_e$ and $x(\xi = +1) = x_{e+1}$, the linear mapping

$$\begin{aligned} \chi_e : \Omega_\square &\to \Omega_e \\ \xi \mapsto x &= \frac{1}{2}[1-\xi]\, x_e + \frac{1}{2}[1+\xi]\, x_{e+1} \end{aligned} \tag{2.34}$$

applies. The ranges for x and ξ in (2.34) are implied by the definitions of $\Omega_\square$ and Ω_e. By means of (2.34), every $\xi \in [-1, 1]$ corresponds to exactly one location in the interval given by x_e and x_{e+1}. Obviously, χ_e has to be invertible, i.e., bijective, for the approach to work. Here, this is implied by linearity and χ_e^{-1} reads

$$\begin{aligned} \chi_e^{-1} : \Omega_e &\to \Omega_\square \\ x \mapsto \xi &= \frac{2x - [x_e + x_{e+1}]}{L_e} \end{aligned}$$

with the length of Ω_e, $L_e = x_{e+1} - x_e$ (Fig. 2.3).

The aim is to approximate the unknown function u by a linear trial function in $\Omega_\square$. A linear function can be expressed by

$$u_\square(\xi) = c_0 + c_1 \xi$$

where $u_\square(\xi)$ refers to the linear approximation in $\Omega_\square$. We denote the values of $u_\square(\xi)$ at $\xi = -1$ and $\xi = 1$ by u_0 and u_1, respectively. Demanding,

$$\begin{aligned} u_\square(\xi = -1) &= u_0 \\ u_\square(\xi = +1) &= u_1 \end{aligned}$$

the coefficients c_1 and c_2 can be expressed in terms of the more meaningful coefficients u_0 and u_1. The final result reads

$$u_\square(\xi) = \frac{1}{2}[1-\xi]\,u_0 + \frac{1}{2}[1+\xi]\,u_1\,. \tag{2.35}$$

Actually, (2.34) is derived by similar arguments.

However, not only the approximation of u is required in the weak form but as well its derivative with respect to x, which involves chain rule due to the fact, that $u_\square$ is expressed in terms of ξ. The derivative of $x(\xi)$ in (2.34) with respect to ξ is the so-called Jacobian j_e of the mapping χ_e,

$$j_e = \frac{\mathrm{d}x}{\mathrm{d}\xi} = -\frac{1}{2}x_e + \frac{1}{2}x_{e+1} = \frac{1}{2}L_e\,.$$

Since, j_e is nothing but the linearisation of χ_e, it depends in general on ξ. Here, j_e is constant because of the linearity of χ_e. The derivative of (2.35) with respect to x is given by

$$u'_\square(\xi) = -\frac{1}{L_e}u_0 + \frac{1}{L_e}u_1\,. \tag{2.36}$$

Using matrix notation, (2.35) and (2.36) can be written compactly as follows

$$u_\square(\xi) = \underline{\boldsymbol{N}}^{\mathrm{T}}\,\underline{\hat{\boldsymbol{u}}}_\square \tag{2.37}$$

$$u'_\square(\xi) = \underline{\boldsymbol{B}}^{\mathrm{T}}\,\underline{\hat{\boldsymbol{u}}}_\square\, j_e^{-1} \tag{2.38}$$

with

$$\underline{\boldsymbol{N}}^{\mathrm{T}} = \left[\tfrac{1}{2}(1-\xi)\;\; \tfrac{1}{2}(1+\xi)\right] \qquad \underline{\boldsymbol{B}}^{\mathrm{T}} = \left[-\tfrac{1}{2}\;\; \tfrac{1}{2}\right]$$

and

$$\underline{\hat{\boldsymbol{u}}}_\square^{\mathrm{T}} = \left[u_0\; u_1\right].$$

Analogously, the test function v and its derivative can be represented by

$$v_\square(\xi) = \underline{\boldsymbol{N}}^{\mathrm{T}}\,\underline{\hat{\boldsymbol{v}}}_\square \tag{2.39}$$

$$v'_\square(\xi) = \underline{\boldsymbol{B}}^{\mathrm{T}}\,\underline{\hat{\boldsymbol{v}}}_\square\, j_e^{-1} \tag{2.40}$$

with

$$\underline{\hat{\boldsymbol{v}}}_\square^{\mathrm{T}} = \left[v_0\; v_1\right]$$

where v_0, v_1 are the values of the test function at the left and right border of the domain $\Omega_\square$, respectively.

Having defined trial function and test function for the reference domain, the next step consists in computing $\mathcal{W}_e$ of (2.33). However, this requires to relate the $\underline{\hat{u}}_\square$ and $\underline{\hat{v}}_\square$ with the corresponding quantities for the domain Ω_e according to Fig. 2.2. There exist various options to achieve this. Here, we use so-called gathering matrices which relate the values of trial and test functions at the left and right border of the reference domain with the corresponding values for a given Ω_e, i.e.,

$$\underline{\hat{u}}_\square \rightarrow \underline{\underline{A}}_e \, \underline{\hat{u}} \, , \qquad \underline{\hat{v}}_\square \rightarrow \underline{\underline{A}}_e \, \underline{\hat{v}} \tag{2.41}$$

with

$$\underline{\hat{u}}^{\mathrm{T}} = \left[u_1 \; u_2 \; \ldots \; u_{N+1} \right]$$
$$\underline{\hat{v}}^{\mathrm{T}} = \left[v_1 \; v_2 \; \ldots \; v_{N+1} \right] .$$

To keep the derivation as transparent as possible, an intermediate step is performed in which a quantity $\mathcal{W}_{e_\square}$ is obtained first which still depends on $\underline{\hat{u}}_\square$ and $\underline{\hat{v}}_\square$. Afterwards, $\mathcal{W}_e$ is derived from $\mathcal{W}_{e_\square}$ by applying (2.41).

Since so far everything is expressed in terms of ξ, the integration should be performed with respect to ξ too. Due to (2.34)

$$\int\limits_{x=x_e}^{x_{e+1}} g(x)\mathrm{d}x = \int\limits_{\xi=-1}^{+1} g(\xi) \, j_e \, \mathrm{d}\xi = \int\limits_{\xi=-1}^{+1} g(\xi) \, \frac{L_e}{2} \, \mathrm{d}\xi \tag{2.42}$$

holds. Performing the integration indicated in (2.33), taking into account (2.37), (2.38), (2.39), and (2.40), yields

$$\mathcal{W}_{e_\square} = \int\limits_{\xi=-1}^{+1} \lambda(\underline{B}^{\mathrm{T}} \, \underline{\hat{u}}_\square)(\underline{B}^{\mathrm{T}} \, \underline{\hat{v}}_\square)\frac{1}{j_e} \mathrm{d}\xi - \int\limits_{\xi=-1}^{+1} n \, \underline{N}^{\mathrm{T}} \, \underline{\hat{v}}_\square \, j_e \, \mathrm{d}\xi \, .$$

Obviously, λ and n have to be expressed as functions of ξ. Furthermore,

$$(\underline{B}^{\mathrm{T}} \, \underline{\hat{u}}_\square)(\underline{B}^{\mathrm{T}} \, \underline{\hat{v}}_\square) = \underline{\hat{v}}_\square^{\mathrm{T}} \, \underline{B} \, \underline{B}^{\mathrm{T}} \, \underline{\hat{u}}_\square$$

holds, which can be seen by simply performing the operations explicitly. In addition, all hatted vectors contain exclusively real numbers and no functions. Therefore, they are not affected by the integration and can be factored out the integral. Hence, $\mathcal{W}_{e_\square}$ can be written concisely as

$$\mathcal{W}_{e_\square} = \underline{\hat{v}}_\square^{\mathrm{T}} \left[\underline{\underline{K}}_e \, \underline{\hat{u}}_\square - \underline{q}_e \right] \tag{2.43}$$

with

$$\underline{\underline{K}}_e = \int_{-1}^{+1} \lambda \, \underline{B} \, \underline{B}^{\mathrm{T}} \frac{1}{j_e} \, \mathrm{d}\xi \tag{2.44}$$

and

$$\underline{q}_e = \int_{-1}^{+1} n \, \underline{N} \, j_e \, \mathrm{d}\xi \, . \tag{2.45}$$

Again, $\underline{\underline{K}}_e$ and $\underline{q}_e$ are known as stiffness matrix and load vector of the element, respectively.

The contribution of an element e to the weak form, $\mathcal{W}_e$, is obtained from (2.43) by applying (2.41), which gives

$$\mathcal{W}_e = \underline{\hat{v}}^{\mathrm{T}} \left[\underline{\underline{A}}_e^{\mathrm{T}} \, \underline{\underline{K}}_e \, \underline{\underline{A}}_e \, \underline{\hat{u}} - \underline{\underline{A}}_e^{\mathrm{T}} \, \underline{q}_e \right] . \tag{2.46}$$

The weak form is finally obtained by introducing (2.46) into (2.32). The final result is

$$\mathcal{W}_h = \underline{\hat{v}}^{\mathrm{T}} \left[\underline{\underline{K}} \, \underline{\hat{u}} - \underline{q} \, \underline{\hat{u}} - \underline{\hat{F}} \right] = 0 \tag{2.47}$$

with

$$\underline{\underline{K}} = \sum_{e=1}^{N} \underline{\underline{A}}_e^{\mathrm{T}} \, \underline{\underline{K}}_e \, \underline{\underline{A}}_e \tag{2.48}$$

and

$$\underline{q} = \sum_{e=1}^{N} \underline{\underline{A}}_e^{\mathrm{T}} \, \underline{q}_e \, . \tag{2.49}$$

Regarding the definition of $\underline{\hat{F}}$, see the discussion below Eq. (2.26).

Because of the arbitrariness of $\underline{\hat{v}}$, (2.47) can only be fulfilled in general if the bracket vanishes, i.e., if

$$\underline{\underline{\mathbf{K}}} \, \underline{\hat{u}} = \underline{q} + \underline{\hat{F}}$$

holds.

If λ is constant, $\lambda(x) = \lambda_0$, the element stiffness matrix $\underline{\underline{K}}_e$ can be computed easily

$$\underline{\underline{K}}_e = \int_{-1}^{+1} \lambda_0 \begin{bmatrix} -\frac{1}{2} \\ +\frac{1}{2} \end{bmatrix} \begin{bmatrix} -\frac{1}{2} & +\frac{1}{2} \end{bmatrix} \frac{1}{j_e} \mathrm{d}\xi = \frac{\lambda_0}{L_e} \begin{bmatrix} 1 & -1 \\ -1 & 1 \end{bmatrix} .$$

Similarly $\underline{q}_e$ for a constant source term, $n(x) = n_0$, is given by

$$\underline{q}_e = n_0 \int_{-1}^{+1} \underline{N}\, j_e \, \mathrm{d}\xi = \frac{n_0}{4} L_e \begin{bmatrix} 2 - (\frac{1}{2} - \frac{1}{2}) \\ 2 + (\frac{1}{2} - \frac{1}{2}) \end{bmatrix} = \frac{n_0 L_e}{2} \begin{bmatrix} 1 \\ 1 \end{bmatrix} .$$

Gathering matrices are boolean matrices because their elements can only be one or zero. Since, the only information encoded by the gathering matrix $\underline{\underline{A}}_e$ is the correspondences $u_0 \rightarrow u_e$ and $u_1 \rightarrow u_{e+1}$, only two entries are different from zero. For a discretization with N intervals or elements, every $\underline{\underline{A}}_e$ has $2 \times (N + 1)$ elements from which only two are different from zero. Therefore, the use of gathering matrices is only recommendable for didactic reasons and small systems. More efficient methods are usually employed for assembling global objects. This is expressed here by means of an assembling operator, which adds the entries of the element stiffness matrix to the global stiffness matrix ensuring the correct correspondences. Analogously, this operator assembles the global load vector. Hence, (2.48) and (2.49) are encoded by means of this operator as follows

$$\underline{\underline{K}} = \mathop{\mathsf{A}}_{e=1}^{N} \underline{\underline{K}}_e , \qquad \underline{q} = \mathop{\mathsf{A}}_{e=1}^{N} \underline{q}_e .$$

In the following, a detailed example is provided. The reader is encouraged to follow carefully the individual steps of the computation.

Example 2.9 The finite element scheme is demonstrated for Example 2.1 with $n = n_0$ and constant stiffness λ_0. Two finite elements with length $L_1 = L_2 = l/2$ are used.

1. *Discretisation*

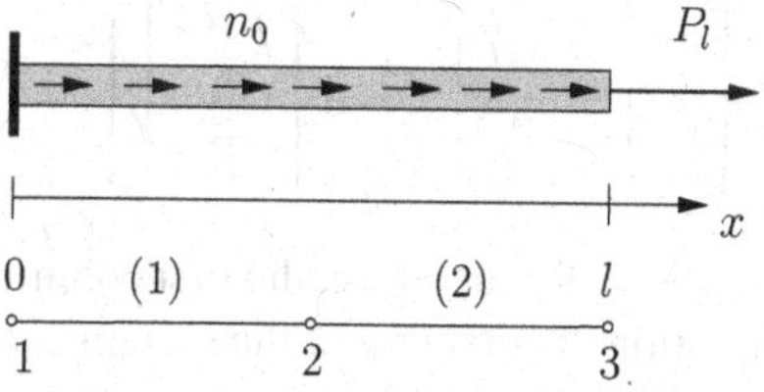

The figure shows the discretisation together with the indexes for the subintervals (finite elements) and the interval borders, also called nodes. The corresponding vector with the unknowns at the nodes is

$$\underline{\hat{u}}^{\mathrm{T}} = (u_1, u_2, u_3) .$$

2. *Assembling element stiffness matrices and load vectors*
Since both elements have same length and stiffness, their stiffness matrices and load vectors are given by

$$\underline{\underline{K}}_1 = \underline{\underline{K}}_2 = 2\frac{\lambda_0}{l}\begin{bmatrix} 1 & -1 \\ -1 & 1 \end{bmatrix}, \quad \underline{q}_1 = \underline{q}_2 = \frac{n_0 l}{4}\begin{bmatrix} 1 \\ 1 \end{bmatrix}.$$

3. *Global stiffness and global load vector*
The relations between Ω_1, Ω_2 and $\Omega_\square$ in terms of nodal values are given by

$$e = 1: \begin{bmatrix} u_0 \\ u_1 \end{bmatrix} = \begin{bmatrix} 1 & 0 & 0 \\ 0 & 1 & 0 \end{bmatrix}\begin{bmatrix} u_1 \\ u_2 \\ u_3 \end{bmatrix} \quad \text{and} \quad e = 2: \begin{bmatrix} u_0 \\ u_1 \end{bmatrix} = \begin{bmatrix} 0 & 1 & 0 \\ 0 & 0 & 1 \end{bmatrix}\begin{bmatrix} u_1 \\ u_2 \\ u_3 \end{bmatrix}$$

from which the gathering matrices $\underline{\underline{A}}_1$ and $\underline{\underline{A}}_2$ can be read off. Global system matrix and global load vector are computed according to

$$\begin{aligned} \underline{\underline{K}} &= \underline{\underline{A}}_1^{\mathrm{T}}\underline{\underline{K}}_1\underline{\underline{A}}_1 + \underline{\underline{A}}_2^{\mathrm{T}}\underline{\underline{K}}_2\underline{\underline{A}}_2 \\ &= \begin{bmatrix} 1 & 0 \\ 0 & 1 \\ 0 & 0 \end{bmatrix}\frac{2\lambda_0}{l}\begin{bmatrix} 1 & -1 \\ -1 & 1 \end{bmatrix}\begin{bmatrix} 1 & 0 & 0 \\ 0 & 1 & 0 \end{bmatrix} + \begin{bmatrix} 0 & 0 \\ 1 & 0 \\ 0 & 1 \end{bmatrix}\frac{2\lambda_0}{l}\begin{bmatrix} 1 & -1 \\ -1 & 1 \end{bmatrix}\begin{bmatrix} 1 & 0 & 0 \\ 0 & 1 & 0 \end{bmatrix} \\ &= \frac{2\lambda_0}{l}\begin{bmatrix} 1 & -1 & 0 \\ -1 & 2 & -1 \\ 0 & -1 & 1 \end{bmatrix} \end{aligned}$$

and

$$\underline{q} = \underline{\underline{A}}_1^{\mathrm{T}}\underline{q}_1 + \underline{\underline{A}}_2^{\mathrm{T}}\underline{q}_2 = \begin{bmatrix} 1 & 0 \\ 0 & 1 \\ 0 & 0 \end{bmatrix}\frac{n_0 l}{4}\begin{bmatrix} 1 \\ 1 \end{bmatrix} + \begin{bmatrix} 0 & 0 \\ 1 & 0 \\ 0 & 1 \end{bmatrix}\frac{n_0 l}{4}\begin{bmatrix} 1 \\ 1 \end{bmatrix} = \frac{n_0 l}{4}\begin{bmatrix} 1 \\ 2 \\ 1 \end{bmatrix}.$$

4. *Global system of equations and essential boundary conditions*
The weak form with our specific choices for trial function and test function, see (2.47), reads in more detail

$$\begin{bmatrix} v_1 & v_2 & v_3 \end{bmatrix}\left[\frac{2\lambda_0}{l}\begin{bmatrix} 1 & -1 & 0 \\ -1 & 2 & -1 \\ 0 & -1 & 1 \end{bmatrix}\begin{bmatrix} u_1 \\ u_2 \\ u_3 \end{bmatrix} - \frac{n_0 l}{4}\begin{bmatrix} 1 \\ 2 \\ 1 \end{bmatrix} - \begin{bmatrix} P(0) \\ 0 \\ P_l \end{bmatrix}\right].$$

The essential boundary condition implies $u_1 = 0$. Therefore, the first column of the stiffness matrix can be deleted. In addition the first row of the system can be deleted too because of $v_1 = 0$. Hence, the first element in $\underline{\hat{F}}$ does not even have to be specified, although according to our knowledge about the boundary value problem, it has to be equal to the support force at $x = 0$. $P(l)$ on the other hand is given explicitly by the natural boundary condition $P(l) = P_l$. The constrained system to be solved for the unknown displacements reads

$$\frac{2\lambda_0}{l}\begin{bmatrix} 2 & -1 \\ -1 & 1 \end{bmatrix}\begin{bmatrix} u_2 \\ u_3 \end{bmatrix} = \frac{n_0 l}{4}\begin{bmatrix} 2 \\ 1 \end{bmatrix} + \begin{bmatrix} 0 \\ P_l \end{bmatrix}.$$

5. *Solution*

$$u_2 = \frac{3}{8}\frac{n_0 l^2}{\lambda_0} + \frac{1}{2}\frac{P_l}{\lambda_0}, \qquad u_3 = \frac{1}{2}\frac{n_0 l^2}{\lambda_0} + \frac{P_l}{\lambda_0}.$$

6. *Accuracy estimation via natural boundary conditions*
The natural boundary condition can be expressed in terms of the computed results as follows

$$\begin{aligned} \lambda_0 u_h'(x = l) &= \frac{\lambda_0}{je}\underline{\boldsymbol{B}}^{\mathrm{T}}\underline{\boldsymbol{A}}_2\,\underline{\hat{\boldsymbol{u}}} = \frac{2\lambda_0}{l}\begin{bmatrix} -\frac{1}{2} & \frac{1}{2} \end{bmatrix}\begin{bmatrix} 0 & 1 & 0 \\ 0 & 0 & 1 \end{bmatrix}\begin{bmatrix} 0 \\ u_2 \\ u_3 \end{bmatrix} \\ &= \frac{\lambda_0}{l}[u_3 - u_2] = \frac{n_0 l}{4} + P_l. \end{aligned}$$

The result coincides with the approximation of the axial force, $\tilde{P}$, in the interval number two. $\tilde{P}$ is necessarily constant because the axial displacements are approximated by linear functions. The result illustrates, that for $n_0 = 0$, the natural boundary condition is exactly fulfilled by the FEM solution because for this case, FEM solution and exact solution coincide. However, for $n_0 \neq 0$, an approximation error can be observed. The latter can be used for choosing the appropriate discretisation according to the required accuracy of the solution.

Apart from more elaborated methods for estimating the error of a finite element solution, accuracy in terms of natural boundary conditions provides already some valuable and easy to access information.

Remark 2.8 Example 2.9 illustrates the different steps of a finite element analysis. The first step is usually performed by the user of a FEM software via some preprocessing tool. Steps 2 to 5 are then executed by the FEM software. The last step however, has to be performed again by the user of a FEM software together with other plausibility checks, which requires sound knowledge about FEM and the boundary value problem.

Remark 2.9 Alternatively to step four in Example 2.9, the system can be handled without restricting the test function. This leads to a system

$$\frac{2\lambda_0}{l}\begin{bmatrix} -1 & 0 & -\frac{l}{2\lambda_0} \\ 2 & -1 & 0 \\ -1 & 1 & 0 \end{bmatrix}\begin{bmatrix} u_2 \\ u_3 \\ P(0) \end{bmatrix} = \frac{n_0 l}{4}\begin{bmatrix} 0 \\ 2 \\ 1 \end{bmatrix} + \begin{bmatrix} 0 \\ 0 \\ P_l \end{bmatrix}$$

which does not only involve more equations than the system derived in step four of Example 2.9. Its condition number also tends to be larger, see Sect. A.8. The condition number κ indicates how noise in the input data affects the accuracy of the solution.

2.1.8 Non-homogeneous Dirichlet Conditions

So far, only homogeneous Dirichlet and Neumann boundary conditions have been considered. Non-homogeneous Dirichlet boundary conditions require some additional effort, which is illustrated in the following by considering the problem

$$\lambda_0 u'' + n_0 = 0 \quad x \in (0, l) \tag{2.50}$$

$$u(0) = 0 \tag{2.51}$$

$$u(l) = \overline{u} \tag{2.52}$$

with constants λ_0, n_0 and $\overline{u}$. Regarding the corresponding weak form, it is clear, that, trial and test functions cannot be in the same function space because the test functions v should vanish at $x = 0$ and $x = l$, whereas the solution should take the value $\overline{u}$ at $x = l$. This is problematic insofar as the Lax-Milgram lemma does not apply in this case, see Sect. 2.1.5 as well as Sect. C.7.

The situation can be remedied by introducing a function U such that

$$U = u - w$$

where u is supposed to be a solution for the original problem. Demanding $w(0) = 0$ and $w(l) = \overline{u}$ implies $U(0) = 0$ and $U(l) = 0$. In consequence, (2.50) takes the form

$$\lambda_0 (U + w)'' + n_0 = 0\,. \tag{2.53}$$

For deriving the corresponding variational formulation of (2.53),

$$\int_0^l w'' v \, \mathrm{d}x = w'v\big|_0^l - \int_0^l w'v' \, \mathrm{d}x = -\int_0^l w'v' \, \mathrm{d}x$$

is required, where the boundary terms vanish because of $v(0) = 0$ and $v(l) = 0$.

Provided, that the function w is known, the task of finding a weak solution can be defined in terms of homogeneous Dirichlet boundary conditions.

Task 2.4 Given $\Omega = (0, l)$, n_0 and λ_0. Find $U = u - w$ such that

$$\lambda_0 \int_0^l U'v'\,dx - n_0 \int_0^l v\,dx + \int_0^l w'\,v'\,dx = 0$$

for $U, v \in H_1^0(\Omega)$ and $w \in H^1(\Omega)$.

Because the boundary value problem (2.53) is linear, it can be written as the superposition of two problems as follows

$$\begin{aligned} \lambda_0 U'' + n_0 &= 0 & w'' &= 0 \\ U(0) &= 0 & w(0) &= 0 \\ U(l) &= 0 & w(l) &= \overline{u} \end{aligned} \tag{2.54}$$

and solving the right hand side problem reveals, that

$$w = \frac{\overline{u}}{l} x\,. \tag{2.55}$$

Regarding the implementation of FEM solution procedures, inhomogeneous Dirichlet boundary conditions are imposed as illustrated in the following using the information provided by Example 2.9. Considering the discretisation used in Example 2.9, the global system of equations reads

$$\frac{2\lambda_0}{l} \begin{bmatrix} 1 & -1 & 0 \\ -1 & 2 & -1 \\ 0 & -1 & 1 \end{bmatrix} \begin{bmatrix} u_1 \\ u_2 \\ u_3 \end{bmatrix} = \frac{n_0 l}{4} \begin{bmatrix} 1 \\ 2 \\ 1 \end{bmatrix} + \begin{bmatrix} P(0) \\ 0 \\ P(l) \end{bmatrix}$$

before taking into account essential boundary conditions. The inhomogeneous Dirichlet condition (2.52) can be incorporated by setting $u_3 = \overline{u}$ and reorganizing the system accordingly. The latter means to rest the product of the third column with $u_3 = \overline{u}$ from both sides, which gives

$$\frac{2\lambda_0}{l} \begin{bmatrix} 1 & -1 & 0 \\ -1 & 2 & 0 \\ 0 & -1 & 0 \end{bmatrix} \begin{bmatrix} 0 \\ u_2 \\ 0 \end{bmatrix} = \frac{n_0 l}{4} \begin{bmatrix} 1 \\ 2 \\ 1 \end{bmatrix} + \overline{u}\,\frac{2\lambda_0}{l} \begin{bmatrix} 0 \\ 1 \\ -1 \end{bmatrix} + \begin{bmatrix} P(0) \\ 0 \\ P(l) \end{bmatrix}$$

by setting in addition $u_1 = 0$ because of (2.51). Furthermore, the first and the last equation have to be deleted since $v_1 = v_2 = 0$. The final results reads

$$u_2 = \frac{n_0 l^2}{4\lambda_0} + \frac{\overline{u}}{2} = U_2 + w_2 ,$$

which indicates, that the discrete version of Task 2.4, i.e., the FEM solution procedure, can be achieved simply by reorganising the global system of equations accordingly, provided, that a function w, which solves the homogeneous problem for prescribed non-homogeneous boundary data, exists. More specifically, only the existence of w has to be ensured but it is not necessary to compute w.

2.1.9 Hand-Calculation Exercises

2.1 Strong form
Solve the problem addressed in Example 2.2 in detail.

2.2 Strong and weak form
Given a rod with constant stiffness λ_0, length l and boundary conditions $u(0) = 0$, $P(l) = F$. Execute the following tasks:

1. Determine the axial displacement u by solving the strong form.
2. Evaluate the left hand side of the weak form $\int_0^l \lambda u' v' \mathrm{d}x - F v(l) = 0$ for u from task 1.
3. Determine the conditions for the test function v under which the weak form holds for arbitrary F.

2.3 Weak form
Solve the boundary value problem (2.5) with $\lambda(x) = \lambda_0$ and homogeneous Dirichlet boundary conditions for

$$n = n_0 \begin{cases} \frac{x}{l} & 0 < x \leq \frac{l}{2} \\ \frac{1}{2} & \frac{l}{2} < x < l \end{cases}$$

by using the variational form.

2.4 Linear algebra
Given:

$$\underline{\underline{A}} = \begin{bmatrix} 1 & 3 \\ 4 & 6 \end{bmatrix} , \quad \boldsymbol{b} = \begin{bmatrix} 10 \\ 22 \end{bmatrix} .$$

Compute $\underline{\underline{A}}\,\underline{b}$, $\underline{b}^{\mathrm{T}}\,\underline{\underline{A}}$ and $\det(\underline{\underline{A}})$.

2.5 Linear algebra

Given $\boldsymbol{a}^{\mathrm{T}} = [1\ 2\ 3]$, $\boldsymbol{b}^{\mathrm{T}} = [1\ x\ x^2]$. Compute $\underline{a}^{\mathrm{T}}\,\underline{b}$, $\int_0^l \boldsymbol{a}^{\mathrm{T}}\,\underline{b}\,\mathrm{d}x$, $\boldsymbol{a}\,\underline{b}^{\mathrm{T}}$ and, $\int_0^l \underline{a}\,\underline{b}^{\mathrm{T}}\,\mathrm{d}x$.

2.6 Linear algebra
Compute the determinant of the matrix

$$\underline{\underline{K}} = \frac{\lambda}{l} \begin{bmatrix} 1 & -1 & 0 \\ -1 & 2 & -1 \\ 0 & -1 & 1 \end{bmatrix} .$$

2.7 Linear algebra
Solve $\underline{\underline{A}}\,\underline{x} = \underline{b}$ for $\underline{x}$ with $\underline{x}^{\mathrm{T}} = [x_1\ x_2]$ and $\underline{\underline{A}}$, $\underline{b}$ from Exercise 2.4.

2.2 Poisson Equation in $\mathbb{R}^N$, $N \geq 2$

2.2.1 *Preliminary Remarks*

Deeper knowledge about the general context and more elaborated math tools are required at the latest when more than one spatial dimension has to be taken into account. Less advanced readers are recommended to examine at least the information provided in Appendices A and B in parallel while proceeding further.

2.2.2 *Strong Form*

The linear partial differential equation

$$-\kappa\, \Delta u = r \tag{2.56}$$

for the unknown scalar field u in a domain $\Omega \subset \mathbb{R}^N$ with boundary $\partial\Omega$ is known as Poisson equation. The corresponding Dirichlet and Neumann boundary conditions read

$$u = \overline{u} \qquad \text{at } \partial\Omega_u\,, \tag{2.57}$$

$$\nabla u \cdot \boldsymbol{n} = \overline{g} \qquad \text{at } \partial\Omega_q\,, \tag{2.58}$$

with $\partial\Omega_u \cup \partial\Omega_q = \partial\Omega$ and the outward unit normal vector $\boldsymbol{n}$ on $\partial\Omega$. The Laplace operator for a Cartesian coordinate system with coordinates x_k reads

$$\Delta = \nabla \cdot \nabla = \frac{\partial^2}{\partial x_i \partial x_i} = \frac{\partial^2}{\partial x_1 \partial x_1} + \frac{\partial^2}{\partial x_2 \partial x_2} + \cdots .$$

More detailed information regarding differential operators and summation convention can be found in Appendix B.

Table 2.2 Some applications covered by the Poisson equation together with the corresponding meaning of involved variables

	Deflection of an elastic membrane	Linear stationary transport of heat or mass	Electrostatics
u	Deflection	Temperature or concentration	Electrostatic potential
$\boldsymbol{q}$	Membrane tension	Heat or mass flux	Electric flux
κ	Stiffness	Heat or mass conductivity	Dielectric constant
r	Out-of-plane pressure	Heat or mass source distribution	Charge density

The homogeneous Poisson equation ($r = 0$) is known as Laplace equation. The Poisson equation has a wide range of applications in science and engineering. A number of examples together with the specific meanings of involved quantities can be found in Table 2.2. Here, stationary heat transport is used for illustration purposes. In this case u is a temperature field, $\boldsymbol{q}$ the heat flux vector, and $\overline{g} = \boldsymbol{q} \cdot \boldsymbol{n}$ denotes the heat flux across the surface. Heat flux and temperature are related by means of the linear constitutive equation

$$\boldsymbol{q} = -\kappa \nabla u \tag{2.59}$$

where κ can even be a second order tensor field. Here, however, only isotropic materials are considered for which κ is either a scalar field or just a constant. The scalar field r represents the spatial distribution of heat sources or sinks.

For a Cartesian coordinate system, the boundary value problem given by (2.56), (2.57), and (2.58), can be written in index notation as follows

$$-\kappa\, u_{,jj} = r \qquad \text{in } \Omega \tag{2.60}$$

$$u = \overline{u} \qquad \text{at } \partial\Omega_u \tag{2.61}$$

$$-\kappa\, u_{,j}\, n_j = \overline{q} \qquad \text{at } \partial\Omega_q \tag{2.62}$$

if summation convention is used as well.

2.2.3 *First Order Weak Partial Derivatives*

Given a real valued function f defined on some domain $\Omega \subset \mathbb{R}^2$, i.e., $f : \Omega \rightarrow \mathbb{R}$. The notation $f(x_1, x_2)$ indicates, that the coordinates x_1 and x_2 are used to address a point in Ω. If x_1 and x_2 are Cartesian coordinates, the first order partial derivative of f with respect to the coordinate x_1 is defined by

$$\frac{\partial f(x_1, x_2)}{\partial x_1} = f_{,1} = \lim_{h \to 0} \frac{1}{h} \left[f(x_1 + h, x_2) - f(x_1, x_2) \right] .$$

The partial derivative of f with respect to x_2 is defined analogously

$$\frac{\partial f(x_1, x_2)}{\partial x_2} = f_{,2} = \lim_{h \to 0} \frac{1}{h} \left[f(x_1, x_2 + h) - f(x_1, x_2) \right] .$$

Partial derivatives are defined locally. They exist in the entire domain Ω, if for every $(x_1, x_2) \in \Omega$ the corresponding limits exist and are unique. The concepts extends in a straight forward manner to $\Omega \subset \mathbb{R}^N$.

An alternative definition is possible by means of Gauss' gradient theorem. Considering two functions f and v, then

$$\int_\Omega \frac{\partial f}{\partial x_1} v \, \mathrm{d}^2 x = \int_\Omega \frac{\partial}{\partial x_1} [f \, v] \, \mathrm{d}^2 x - \int_\Omega f \frac{\partial v}{\partial x_1} \, \mathrm{d}^2 x$$

provided, that the strong partial derivatives exist for both functions. Applying (B.19) to the first term of the right hand side gives

$$\int_\Omega \frac{\partial f}{\partial x_1} v \, \mathrm{d}^2 x = \int_{\partial\Omega} [f \, v] \, n_1 \, \mathrm{d}^2 x - \int_\Omega f \frac{\partial v}{\partial x_1} \, \mathrm{d}^2 x$$

which simplifies to

$$\int_\Omega \frac{\partial f}{\partial x_1} v \, \mathrm{d}^2 x = - \int_\Omega f \frac{\partial v}{\partial x_1} \, \mathrm{d}^2 x$$

if the function v vanishes everywhere on the boundary $\partial\Omega$. This scheme extends to arbitrary dimensions and it can be applied to all coordinates involved. Using the short hand notation for partial derivatives, the general form

$$\int_\Omega f_{,i} \, v \, \mathrm{d}^N x = - \int_\Omega f \, v_{,i} \, \mathrm{d}^N x \tag{2.63}$$

is obtained for $v = 0$ on $\partial\Omega$. As long as the first order partial derivatives of f and v exist everywhere in Ω, (2.63) is just an identity. However, an independent meaning can be assigned to the right hand side of (2.63) by considering

$$\int_\Omega g_i \, v \, \mathrm{d}^N x = - \int_\Omega f \, v_{,i} \, \mathrm{d}^N x \tag{2.64}$$

and interpreting (2.64) as follows. If there exists a function g_i with $i = 1, \ldots, N$ which fulfils (2.64) for a given f for each continuous function v which vanishes on $\partial\Omega$ then g_i is called the weak first order partial derivative with respect to the i-th coordinate. The character of the derivative can be made explicit by notation, e.g., $D_i := g_i$. Often, however, no notational distinction is made as long as the meaning is clear from the context.

Higher order weak partial derivatives can be defined analogously, which is omitted here because they are not required for the problems discussed within this book. As already mentioned in Sect. 2.1.5, a number of important technical details are not discussed in the main part of the book. The interested reader is referred to Appendix C where more detailed information is provided.

2.2.4 *Variational and Weak Form*

In order to derive the variational form of the boundary value problem, (2.60) is multiplied by a test function v and the product is integrated with respect to Ω

$$\int\limits_{\Omega} \left[\kappa u_{,jj} + r\right] v \,\mathrm{d}^N x = 0\,. \tag{2.65}$$

Due to the product rule for differentiation, the first term of the left hand side can be written as

$$\int\limits_{\Omega} \kappa\, u_{,jj}\, v \,\mathrm{d}^N x = \int\limits_{\Omega} \kappa\, [u_{,j}\, v]_{,j} \,\mathrm{d}^N x - \int\limits_{\Omega} \kappa\, u_{,j}\, v_{,j} \,\mathrm{d}^N x\,.$$

Applying Gauss integration theorem (see Appendix B) to the first term of the right hand side yields

$$\int\limits_{\Omega} \kappa\, u_{,jj}\, v \,\mathrm{d}^N x = \int\limits_{\partial\Omega} \kappa\, u_{,j}\, v \,\mathrm{d}S - \int\limits_{\Omega} \kappa\, u_{,j}\, v_{,j} \,\mathrm{d}^N x\,.$$

By means of this result, Eq. (2.65) can be written as

$$\int\limits_{\Omega} \kappa\, u_{,j}\, v_{,j} \,\mathrm{d}^N x - \int\limits_{\Omega} r\, v \,\mathrm{d}^N x - \int\limits_{\partial\Omega_u} \kappa\, u_{,j}\, v \,\mathrm{d}S - \int\limits_{\partial\Omega_q} \kappa\, u_{,j}\, v \,\mathrm{d}S = 0 \tag{2.66}$$

since $\Omega = \Omega_u \bigcup \Omega_q$. Inspection of equation (2.66) reveals, that the boundary condition (2.62) can be plugged directly into the most right hand side term. Furthermore, considering only test functions v which vanish at the boundary $\partial\Omega_u$ yields eventually the variational form of the boundary value problem

$$\mathcal{W} = \int_{\Omega} \kappa\, u_{,j}\, v_{,j}\, \mathrm{d}^N x - \int_{\Omega} r\, v\, \mathrm{d}^N x + \int_{\partial\Omega_q} \overline{q}\, v\, \mathrm{d}S = 0 \tag{2.67}$$

$$u = \overline{u} \quad \text{on} \quad \partial\Omega_u \tag{2.68}$$

and the weak form is obtained from it by interpreting derivatives in the weak sense.

A suitable test function space W consists of all functions with first-order weak derivative in Ω which vanish at $\partial\Omega_u$. Therefore,

$$W(\Omega) = \{f \mid f \in H^1(\Omega) \text{ and } f = 0 \text{ on } \partial\Omega_u\}\,. \tag{2.69}$$

is defined. Trial functions, on the other hand, must fulfil the Dirichlet boundary conditions (2.68), which are not necessarily homogeneous. Therefore, the space

$$V(\Omega) = \{g \mid g \in H^1(\Omega) \text{ and } g = \overline{u} \text{ on } \partial\Omega_u\}\,. \tag{2.70}$$

is defined as trial function space. The corresponding task can be stated as follows.

Task 2.5 Find $u \in V(\Omega)$ according to (2.70) such that

$$\mathcal{W} = \int_{\Omega} \kappa\, u_{,j}\, v_{,j}\, \mathrm{d}^N x - \int_{\Omega} r\, v\, \mathrm{d}^N x + \int_{\partial\Omega_q} \overline{q}\, v\, \mathrm{d}S = 0$$

for arbitrary $v \in W$ defined by (2.69) and given $\kappa, r \in L^2(\Omega), \overline{q} \in L^2(\partial\Omega)$.

If homogeneous Dirichlet boundary conditions are prescribed, i.e., $\overline{u} = 0$ in (2.68), the spaces V and W coincide and the Lax-Milgram lemma applies. Otherwise, analogously to Sect. 2.1.8, Task 2.5 can be reformulated for homogeneous Dirichlet boundary conditions by defining $U = u - w$ with $w \in H^1(\Omega)$. Since $\Omega \in \mathbb{R}^N$, assigning boundary values involves the so-called trace theorem, see Sect. C.9. The reformulation of Task 2.5 is omitted here, because it is equivalent to a reorganisation of the global system of equations in the FEM solution procedure as discussed in Sect. 2.1.8.

2.2.5 *Galerkin FEM with Linear Quadrilateral Elements*

The first step in the process of solving the weak form numerically consists in defining a discretisation of the domain Ω, i.e., dividing Ω into sub-domains Ω_e such that

$$\Omega \approx \bigcup_{e=1}^{N_e} \Omega_e$$

where N_e is the number of sub-domains. Such a discretization is usually called a mesh. A mesh is just a tessellation of a domain $\Omega \subset \mathbb{R}^N$ and as such a geometric entity. It becomes a finite element mesh after defining the physics on it in terms of trial functions, test functions, etc.

Modifications of Task 2.5, similar to those discussed in Sect. 2.1.6, yield its discrete counterpart. This step is not performed here explicitly but recommended to the reader as an exercise.

For a given finite element mesh, the discrete version of Eq. (2.67) can be evaluated by summing up the contributions of the individual domains, i.e.,

$$\mathcal{W}_h = \sum_{e=1}^{N_e} \mathcal{W}_e \,. \tag{2.71}$$

The procedure is first illustrated for $\Omega \subset \mathbb{R}^2$ using linear quadrilateral elements. An example is shown in Fig. 2.4 together with the corresponding reference domain. The latter is a square with a coordinate system defined by the coordinates $\xi_1, \xi_2 \in [-1, 1]$ as illustrated in Fig. 2.4. $\mathcal{W}_e$ in (2.71) takes the form

$$\mathcal{W}_e = \int_{\Omega_e} \kappa\, u_{,j}\, v_{,j}\, \mathrm{d}^2x - \int_{\Omega_e} r\, v\, \mathrm{d}^2x + \int_{\partial\Omega_{e(q)}} \overline{q}\, v\, \mathrm{d}S \,. \tag{2.72}$$

Since $\Omega \subset \mathbb{R}^2$, indexes take only the values 1 and 2.

Reference domain $\Omega_\square$ and Ω_e are related by a mapping χ_e, which is actually a coordinate transform given by

$$\begin{aligned} \chi_e :& \Omega_\square \to \Omega_e \\ &(\xi_1, \xi_2) \mapsto \left(\chi_{e_1} = \sum_{\alpha=1}^{4} N_\alpha(\xi_1, \xi_2)\, x_{1\alpha} \,,\ \chi_{e_2} = \sum_{\alpha=1}^{4} N_\alpha(\xi_1, \xi_2)\, x_{2\alpha} \right) . \end{aligned} \tag{2.73}$$

The functions $N_\alpha(\xi_1, \xi_2)$

$$N_\alpha = \frac{1}{4}\, [1 + a_\alpha \xi_1]\, [1 + b_\alpha \xi_2] \qquad \alpha = 1, ..., 4 \tag{2.74}$$

with coefficients $\{a_\alpha\} = \{-1, 1, -1, 1\}$ and $\{b_\alpha\} = \{-1, -1, 1, 1\}$ are the result of the following procedure. A domain Ω_e is completely defined by four nodes which are connected by straight lines. Every node α can be identified by its position with respect to a coordinate system with coordinates x_1 and x_2. More specifically, a node α has coordinates $(x_{1\alpha}, x_{2\alpha})$. For every coordinate, for instance x_1, there are exactly

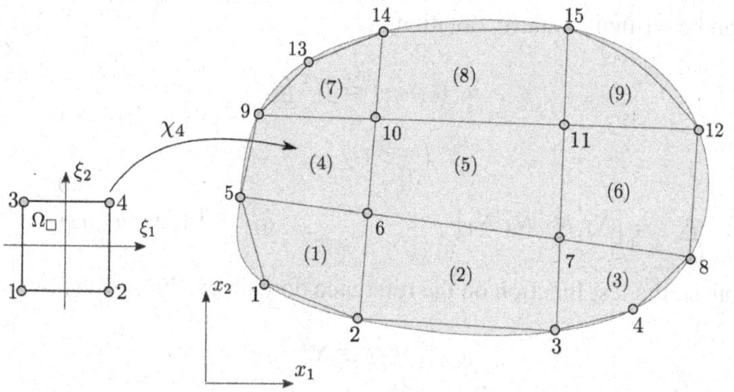

Fig. 2.4 Example discretisation of a two-dimensional domain Ω using quadrilateral sub-domains

four independent pieces of information due to the four nodes. The mapping χ_e is the solution of the corresponding interpolation problem and it returns the coordinates with respect to the $x_1 - x_2$ coordinate system for given $\xi_1, \xi_2 \in [-1, 1]$.

After identifying a particular node of a domain Ω_e with node number one of the reference domain, the remaining correspondences are fixed.

Example 2.10 Suppose, the coordinates of the nodes 5, 6, 9 and 10 in Fig. 2.4 are given as follows

node	5	6	9	10
x_1	0.8	4.9	1.0	5.0
x_2	5.0	4.8	10.0	9.0

and nodes 5, 6, 9 and 10 are associated with the nodes 1, 2, 3 and 4 of $\Omega_\square$. Then χ_{41} of the mapping χ_4 reads

$$\begin{aligned}\chi_{41} = \;& \frac{1}{4}[1-\xi_1][1-\xi_2]\,0.8 + \frac{1}{4}[1+\xi_1][1-\xi_2]\,4.9 \\ &+ \frac{1}{4}[1-\xi_1][1+\xi_2]\,1.0 + \frac{1}{4}[1+\xi_1][1+\xi_2]\,5.0\,,\end{aligned}$$

by which the x_1 coordinate of every point in Ω_4 can be computed by choosing ξ_1 and ξ_2 from the interval $[-1, 1]$. The part χ_{42} is left as an exercise to the reader.

The trial function $u_\square$ on the reference domain $\Omega_\square$ can be expressed as well by means of the shape functions together with the values of the trial function at the nodes, u_α, as follows

$$u_\square(\xi_1, \xi_2) = \sum_{\alpha=1}^{4} N_\alpha(\xi_1, \xi_2) u_\alpha$$

which can be written in matrix notation as

$$u_\square(\xi_1, \xi_2) = \underline{\boldsymbol{N}}^\mathrm{T}\, \hat{\underline{\boldsymbol{u}}}_\square$$

with

$$\underline{\boldsymbol{N}}^\mathrm{T} = \left[N_1\ N_2\ N_3\ N_4\right], \qquad \hat{\underline{\boldsymbol{u}}}_\square^\mathrm{T} = \left[u_1\ u_2\ u_3\ u_4\right].$$

Analogously, the test function on the reference domain can be chosen as

$$v_\square(\xi_1, \xi_2) = \underline{\boldsymbol{N}}^\mathrm{T}\, \hat{\underline{\boldsymbol{v}}}_\square$$

with

$$\hat{\underline{\boldsymbol{v}}}_\square^\mathrm{T} = \left[v_1\ v_2\ v_3\ v_4\right],$$

where the v_i are the values of the test function at the four nodes of the reference domain. Since (2.66) contains the gradients of u and v, we define corresponding objects using matrix notation

$$\nabla_{\underline{x}} u_\square = \begin{bmatrix} \frac{\partial}{\partial x_1} u_\square(\xi_1, \xi_2) \\ \frac{\partial}{\partial x_2} u_\square(\xi_1, \xi_2) \end{bmatrix} = \underline{\underline{\boldsymbol{B}}}\, \hat{\underline{\boldsymbol{u}}}_\square, \qquad \nabla_{\underline{x}} v_\square = \begin{bmatrix} \frac{\partial}{\partial x_1} v_\square(\xi_1, \xi_2) \\ \frac{\partial}{\partial x_2} v_\square(\xi_1, \xi_2) \end{bmatrix} = \underline{\underline{\boldsymbol{B}}}\, \hat{\underline{\boldsymbol{v}}}_\square$$

with

$$\underline{\underline{\boldsymbol{B}}} = \left[\underline{\boldsymbol{B}}_1\ \underline{\boldsymbol{B}}_2\ \underline{\boldsymbol{B}}_3\ \underline{\boldsymbol{B}}_4\right]$$

and

$$\underline{\boldsymbol{B}}_\alpha^\mathrm{T} = \left[\frac{\partial N_\alpha}{\partial x_1}\ \frac{\partial N_\alpha}{\partial x_2}\right].$$

With the information provided so far, the contribution of a finite element e to the weak form (2.71) expressed using the reference domain, $\mathcal{W}_{e_\square}$, can be written as

$$\mathcal{W}_{e_\square} = \hat{\underline{\boldsymbol{v}}}_\square^\mathrm{T} \int\limits_{\Omega_e} \kappa\, \underline{\underline{\boldsymbol{B}}}^\mathrm{T}\, \underline{\underline{\boldsymbol{B}}}\ \mathrm{d}^2x\, \hat{\underline{\boldsymbol{u}}}_\square - \hat{\underline{\boldsymbol{v}}}_\square^\mathrm{T} \int\limits_{\Omega_e} r\, \underline{\boldsymbol{N}}\, \mathrm{d}^2x - \hat{\underline{\boldsymbol{v}}}_\square^\mathrm{T} \int\limits_{\partial\Omega_e} \underline{\boldsymbol{N}}\, \overline{q}\, \mathrm{d}S$$

or, adapting matrix notation

$$\mathcal{W}_{e_\square} = \hat{\underline{\boldsymbol{v}}}_\square^\mathrm{T} \left[\underline{\underline{\boldsymbol{K}}}_e\, \hat{\underline{\boldsymbol{u}}}_\square - \underline{\boldsymbol{r}}_e - \underline{\boldsymbol{q}}_e\right] \tag{2.75}$$

with

$$\underline{\underline{K}}_e = \int_{\Omega_e} \kappa\, \underline{\underline{B}}^{\mathrm{T}}\, \underline{\underline{B}}\, \mathrm{d}^2x\,, \qquad \underline{r}_e = \int_{\Omega_e} r\, \underline{N}\, \mathrm{d}^2x\,, \qquad \underline{q}_e = \int_{\partial\Omega_e} \underline{N}\, \overline{q}\, \mathrm{d}S\,. \tag{2.76}$$

Eventually, the aim is to integrate with respect to the reference domain $\Omega_{\square}$. This, however, requires some preliminary considerations which are more elaborated compared with the spatially one-dimensional case. This issue will be dealt with in more detail in Sect. 2.3.5.

Finally, (2.71) can be written concisely as

$$\mathcal{W}_h = \underline{\hat{v}}^{\mathrm{T}} \left[\underline{\underline{K}}\,\underline{\hat{u}} - \underline{r} - \underline{q} \right] = 0 \tag{2.77}$$

with

$$\underline{\underline{K}} = \mathop{\mathsf{A}}_{e=1}^{N_e} \underline{\underline{K}}_e\,, \qquad \underline{r} = \mathop{\mathsf{A}}_{e=1}^{N_e} \underline{r}_e\,, \qquad \underline{q} = \mathop{\mathsf{A}}_{e=1}^{N_e} \underline{q}_e\,.$$

As before, the assembling operator A adds the element contributions to the global system, ensuring the correct correspondences.

Due to the arbitrariness of the test function values at the nodes, it follows from (2.77), that

$$\underline{\underline{K}}\,\underline{\hat{u}} = \underline{r} + \underline{q}\,. \tag{2.78}$$

After incorporating essential boundary conditions, the system of linear equations (2.78) can be solved to determine the unknown vector $\underline{\hat{u}}$. Afterwards, derived quantities such as the flux vector can be computed by corresponding postprocessing routines.

2.3 Systematic Construction of Lagrange Elements

2.3.1 *Overview*

Recall, that the space of all possible solutions of a given boundary value problem, called V, is infinite dimensional. The idea to search for a solution of a boundary value problem in a finite dimensional subspace of V is common to all numerical methods for solving boundary value problems numerically. In the context of FEM, this finite dimensional subspace is commonly called V_h.

The Finite-Element solution schemes discussed in Sects. 2.1.6 and 2.2.5 illustrate, that $V_h(\Omega)$ requires a mesh. The latter is a discretisation of the domain Ω in terms of entities such as nodes, edges, faces and cells. Relations between nodes and edges, i.e., which nodes define a particular edge, edges and faces, as well as faces and cells determine the topology of the mesh. Its geometry is determined by the coordinates of the nodes with respect to a coordinate system. The mesh approximates the domain Ω, i.e.,

$$\Omega \approx \bigcup_{e=1}^{N} \Omega_e ,$$

where the sub domains Ω_e correspond to edges, faces or cells depending on whether Ω is a spatially one, two or three-dimensional domain, respectively. Since a mesh is a discrete version of a continuous geometric object, many interrelations with other disciplines such as Computer Graphics or Stochastic Geometry exist.

Furthermore, defining trial and test functions as elements of $V_h(\Omega)$ relies on the interpolation of functions based on their values at the nodes of the corresponding mesh. The interpolation scheme can be developed in terms of a global basis $\{\varphi_e(x)\}$ which must be compatible with the mesh. Furthermore, the basis defines interpolation order and continuity properties.

Alternatively, a reference domain $\Omega_\square$ can be defined together with bijective mappings, $\chi_e : \Omega_\square \to \Omega_e$, required for establishing the relations between sub domains and reference domain. Interpolation is performed using the reference domain together with a local interpolation basis $\{N_i\}$. The corresponding global basis can be constructed uniquely from the local basis. Here, only Lagrange polynomial interpolation is discussed.

Constructing the mappings χ_e leads to an interpolation problem as well and the corresponding basis is denoted by $\{S_j\}$. Since the χ_e determine the shape of a domain Ω_e, the functions S_j are called shape functions. Finite elements are called subparametric, isoparametric or superparametric, depending on whether the polynomial order of shape functions is lower, equal or higher than the polynomial order of the local interpolation basis. Within this section only isoparametric elements are considered. An example for a subparametric element can be found in Sect. 2.5 in the context of fourth-order boundary value problems.

A finite element space using a reference domain approach consists of the following ingredients:

- mesh,
- reference domain $\Omega_\square$,
- interpolation basis $\{N_i\}$ defined for $\Omega_\square$,
- bijective mappings $\chi_e : \Omega_\square \to \Omega_e$ defined via shape functions $\{S_i\}$.

In the remaining part of this section, the systematic construction of families of Lagrange elements based on Lagrange interpolation in $\mathbb{R}$ is discussed in detail for quadrilateral and triangular reference domains. Eventually, determining stiffness matrices and load vectors requires integration, preferably performed with respect to the reference domain as discussed at the end of this section.

2.3.2 *Lagrange Interpolation in* $\mathbb{R}$

In the following, the space of all polynomials of degree k defined over a reference domain $\Omega_\square$ is denoted by $\mathcal{P}^k(\Omega_\square)$. The degree k indicates the largest exponent in a monomial. For instance

$$p_1(x) = c_0 + c_1\xi \tag{2.79}$$

with $c_0, c_1 \in \mathbb{R}$ is an element of $\mathcal{P}^1(\Omega_\square)$. The latter is a vector space of dimension two. A possible basis of $\mathcal{P}^1(\Omega_\square)$ is $\{1, \xi\}$. This scheme extends in a straight-forward manner for higher k. A more efficient basis is however the so-called Lagrange basis

$$\mathcal{L}_m^k(\xi) = \prod_{\substack{i=0 \\ i \neq m}}^{k} \frac{\xi - \xi_m}{\xi_i - \xi_m} \quad . \tag{2.80}$$

since it is already designed for interpolation problems with given values at $\xi = \xi_m$, $m = 0, ..., k$. Furthermore, it can be seen from (2.80), that Lagrange polynomials are designed such that $\mathcal{L}_m^k(\xi = \xi_i) = 1$ and $\mathcal{L}_m^k(\xi = \xi_m) = 0$, i.e.,

$$\mathcal{L}_m^k(\xi = \xi_i) = \delta_{mi} , \tag{2.81}$$

where δ_{mi} is the Kronecker-symbol (see, Definition A.17).

The case $k = 1$ with $\xi_0 = -1$ and $\xi_1 = 1$ has been considered within the previous section. The corresponding Lagrange basis elements are

$$\mathcal{L}_0^1(\xi) = \frac{1}{2}[1 - \xi], \qquad \mathcal{L}_1^1(\xi) = \frac{1}{2}[1 + \xi] \tag{2.82}$$

and the solution of the linear interpolation problem for a function $f(\xi)$ with values f_0 and f_1 at $\xi_0 = -1$ and $\xi_1 = 1$, respectively, reads

$$f(\xi) = f_0\, \mathcal{L}_0^1(\xi) + f_1\, \mathcal{L}_1^1(\xi) . \tag{2.83}$$

Not surprisingly, $\mathcal{L}_0^1(\xi)$ and $\mathcal{L}_1^1(\xi)$ coincide with the functions $N_1(\xi)$ and $N_2(\xi)$, respectively, used in Sect. 2.1.7 for trial and test function, see, e.g., (2.35). Lagrange basis functions in one dimension are given in Table 2.3 for different interpolation orders k up to $k = 3$.

Table 2.3 One dimensional Lagrange polynomials for different interpolation orders k up to $k = 3$

k	Graphs	Coordinates and shape functions
1	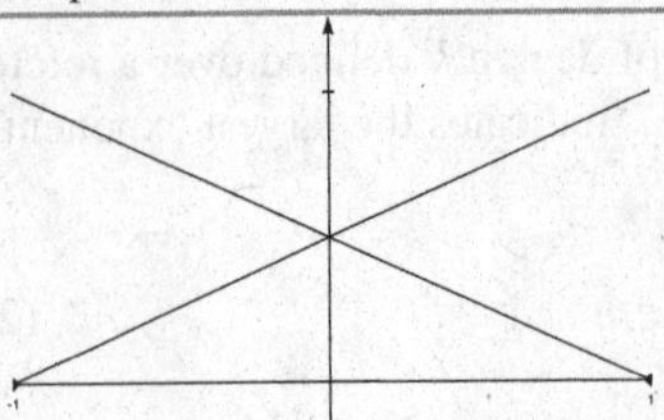	• $\xi_0 = -1$ • $\xi_1 = 1$ • $\mathcal{L}_0^1 = \frac{1}{2}[1-\xi]$ • $\mathcal{L}_1^1 = \frac{1}{2}[1+\xi]$
2	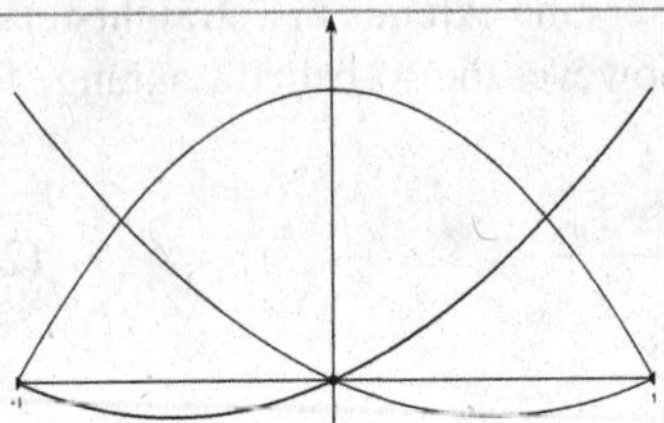	• $\xi_0 = -1$ • $\xi_1 = 0$ • $\xi_2 = 1$ • $\mathcal{L}_0^2 = \frac{1}{2}\xi[\xi-1]$ • $\mathcal{L}_1^2 = [1+\xi][1-\xi]$ • $\mathcal{L}_2^2 = \frac{1}{2}\xi[\xi+1]$
3	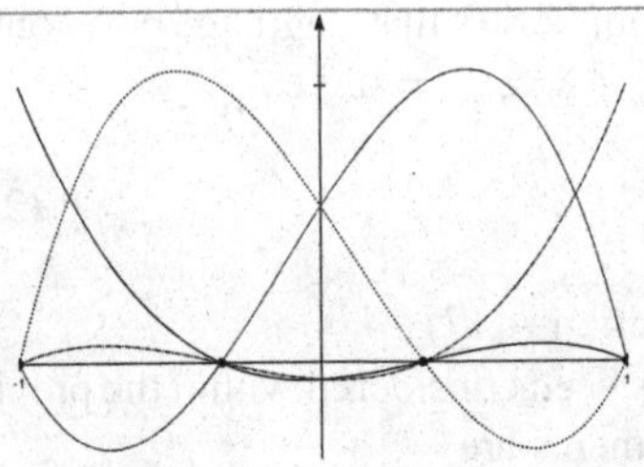	• $\xi_0 = -1$ • $\xi_1 = -\frac{1}{3}$ • $\xi_2 = \frac{1}{3}$ • $\xi_3 = 1$ • $\mathcal{L}_0^3 = -\frac{9}{16}[\xi+\frac{1}{3}][\xi-\frac{1}{3}][\xi-1]$ • $\mathcal{L}_1^3 = \frac{27}{16}[\xi+1][\xi-\frac{1}{3}][\xi-1]$ • $\mathcal{L}_2^3 = -\frac{27}{16}[\xi+1][\xi+\frac{1}{3}][\xi-1]$ • $\mathcal{L}_3^3 = \frac{9}{16}[\xi+1][\xi+\frac{1}{3}][\xi-\frac{1}{3}]$

2.3.3 *Q-type Interpolation Bases in* $\mathbb{R}^N, N \geq 2$

The linear quadrilateral finite element used in the previous section is a particular representative of a family of finite elements known as Lagrange elements with quadrilateral reference domain. These elements are based on Lagrange interpolation which allows for a systematic and straightforward construction of such finite elements in terms of spatial dimension and interpolation order.

Recall $\mathcal{P}^k(\Omega_\square)$, i.e, the space of all polynomials of degree $d \leq k$ with respect to a domain $\Omega_\square$ from Sect. 2.3.2. $\mathcal{P}^k(\Omega_\square)$ is a vector space and a suitable basis can be constructed using Lagrange polynomials.

A reference domain for the two-dimensional case can be constructed by means of the Cartesian product $\Omega_\square \times \Omega_\square = \Omega_\square^2$. Only in one dimension, a simple degree is sufficient for characterizing polynomials. In higher dimensions, a total degree d_t and a partial degree d_p are necessary. The former refers to the highest sum of exponents in a monomial whereas the latter is just the highest exponent of an individual variable.

Polynomial degrees

The polynomial

$$p(\xi_1, \xi_2) = c_{00} + c_{10}\xi_1 + c_{01}\xi_2 + c_{11}\xi_1\xi_2$$

with real coefficients c_{ij} and coordinates ξ_1 and ξ_2 has partial degree one and total degree two.

In the following, the spaces of all polynomials of two variables with total degree $d_t \leq k$ and partial degree $d_p \leq k$ defined for $\Omega_\square$ are called $\mathcal{P}^k(\Omega_\square^2)$ and $\mathcal{Q}^k(\Omega_\square^2)$, respectively. Given two polynomials

$$p(\xi_1) = A + B\xi_1 \,,$$
$$q(\xi_2) = C + D\xi_2 \,,$$

with real coefficients A, B, C, D and coordinates ξ_1 and ξ_2. Their product

$$p(\xi_1)q(\xi_2) = A\,C + C\,B\,\xi_1 + A\,D\,\xi_2 + B\,D\,\xi_1\xi_2 \tag{2.84}$$

is an element of the vector space $\mathcal{Q}^1(\Omega_\square^2)$. It turns out, that the multiplication (2.84) has all properties of a tensor product, see Appendix A. More specifically, $\mathcal{Q}^1(\Omega_\square^2)$ is a tensor product space

$$\mathcal{Q}^1(\Omega_\square^2) = \mathcal{P}^1(\Omega_\square) \otimes \mathcal{P}^1(\Omega_\square) = \otimes^2\mathcal{P}^1(\Omega_\square)$$

provided that the individual $\mathcal{P}^1(\Omega_\square)$ refer to different coordinates. A basis for $\mathcal{Q}^1(\Omega_\square^2)$ can easily be constructed by the tensor product of the individual sets of base vectors.

Basis of $\mathcal{Q}^1(\Omega_\square^2)$

Considering two spaces $\mathcal{P}^1(\Omega_\square)$ with the following sets of base vectors

$$\{g_1, g_2\} = \left\{\frac{1}{2}[1 - \xi_1], \frac{1}{2}[1 + \xi_1]\right\} \,,$$
$$\{\hat{g}_1, \hat{g}_2\} = \left\{\frac{1}{2}[1 - \xi_2], \frac{1}{2}[1 + \xi_2]\right\} \,.$$

A basis for $\mathcal{Q}^1(\Omega_\square^2)$ is constructed by

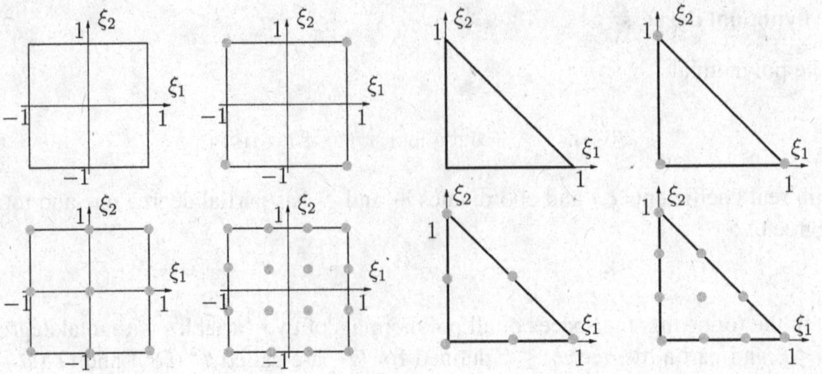

Fig. 2.5 Unit square (left) and unit triangle (right) as reference domains together with corresponding Lagrange elements of different interpolation order

$$N_1 = \boldsymbol{g}_1 \otimes \hat{\boldsymbol{g}}_1 = \frac{1}{4}[1-\xi_1][1-\xi_2], \qquad N_3 = \boldsymbol{g}_1 \otimes \hat{\boldsymbol{g}}_2 = \frac{1}{4}[1-\xi_1][1+\xi_2],$$

$$N_2 = \boldsymbol{g}_2 \otimes \hat{\boldsymbol{g}}_1 = \frac{1}{4}[1+\xi_1][1-\xi_2], \qquad N_4 = \boldsymbol{g}_2 \otimes \hat{\boldsymbol{g}}_2 = \frac{1}{4}[1+\xi_1][1+\xi_2],$$

and the reader is encouraged to compare this result with the functions $N_\alpha(\xi_1, \xi_2)$ (see, (2.74)) of the previous section.

The graphs of basis functions of the previous example are shown in Fig. 2.6. The corresponding finite element is known as linear quadrilateral Lagrange element. This scheme extends in a straight-forward manner towards higher dimensions as well as higher polynomial order and allows to construct families of Lagrange elements for N-dimensional squares based on

$$\mathcal{Q}^k(\Omega_\square^N) = \otimes^N \mathcal{P}^k(\Omega_\square)$$

with polynomial order k and spatial dimension N. The dimension of polynomial spaces constructed in this way is given by

$$\dim\left(\mathcal{Q}^k(\Omega_\square^N)\right) = [k+1]^N .$$

Reference domains with corresponding interpolation nodes are shown in Fig. 2.5 for quadrilateral Lagrange elements up to interpolation order three. Constructing the sets of basis functions for quadrilateral quadratic and cubic Lagrange elements based on the information given in Table 2.3 and the scheme discussed above is left as an exercise to the reader.

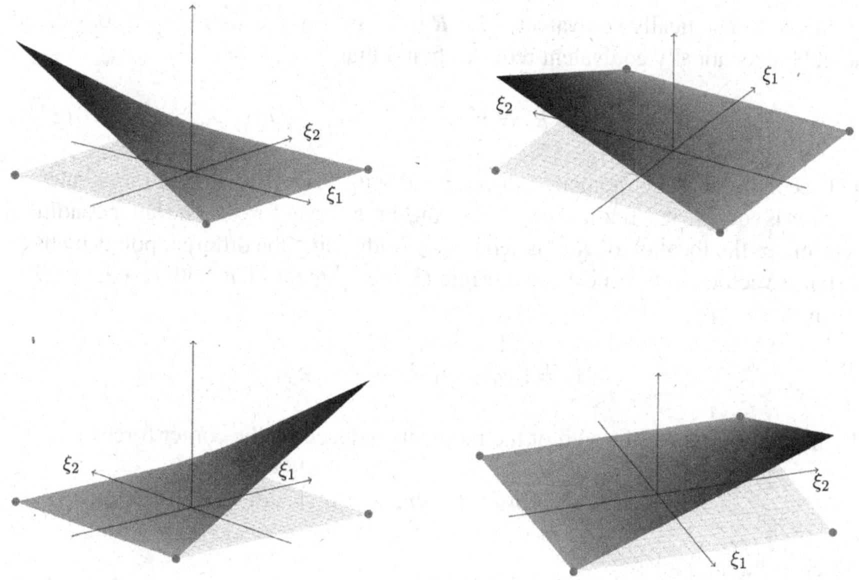

Fig. 2.6 Shape functions for the linear quadrilateral element

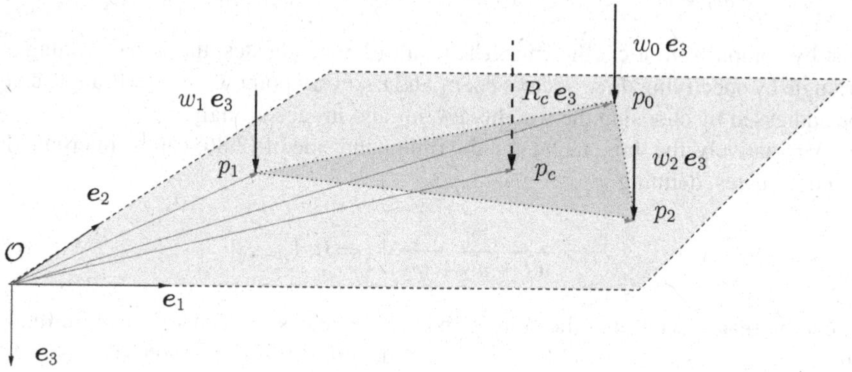

Fig. 2.7 Illustration of barycentric coordiantes by means of the center of gravity of a triangular region with weights centered at the corner points

2.3.4 P-type Interpolation Bases in $\mathbb{R}^N$, $N \geq 2$

For geometric simplexes, such as triangles and their higher-dimensional analogues, a systematic construction of Lagrange elements is possible as well. Before proceeding, it is convenient to discuss barycentric coordinates first.

As the name already indicates, barycentric coordinates emerge naturally in the context of determining the center of gravity of a triangular region with weights concentrated at the corner points as shown in Fig. 2.7. The three nodal forces can be replaced by a statically equivalent force $\boldsymbol{R} = R_c\,\boldsymbol{e}_3$. For the simple case considered here, being statically equivalent requires firstly that

$$R_c = w_0 + w_1 + w_2\,, \tag{2.85}$$

and secondly, that the moment induced by $\boldsymbol{R}$ with respect to an arbitrary center of rotation is equal to the moment induced by the three corner forces. The latter condition determines the location of $\boldsymbol{R}$ denoted by p_c. Addressing the different points by their distance vectors with respect to an origin O, the moment of $\boldsymbol{R}$ with respect to O is given by

$$M_r = [w_0 + w_1 + w_2]\boldsymbol{r}_c \wedge \boldsymbol{e}_3\,.$$

M_r has to be equal to the sum of the moments induced by the corner forces

$$M_{012} = [w_0\boldsymbol{r}_0 + w_1\boldsymbol{r}_1 + w_2\boldsymbol{r}_2] \wedge \boldsymbol{e}_3$$

which eventually gives

$$\boldsymbol{r}_c = \frac{w_0}{w_0 + w_1 + w_2}\boldsymbol{r}_0 + \frac{w_1}{w_0 + w_1 + w_2}\boldsymbol{r}_1 + \frac{w_2}{w_0 + w_1 + w_2}\boldsymbol{r}_2 \tag{2.86}$$

just by comparison of coefficients. The result (2.86) indicates, that after defining a triangle by specifying three vectors $\boldsymbol{r}_0$, $\boldsymbol{r}_1$ and $\boldsymbol{r}_2$, every point within the triangle can be addressed by choosing the weights w_0, w_1 and w_2 accordingly.

Alternatively, the three factors on the right hand side of (2.86) can be interpreted as coordinates, defining

$$\lambda_i = \frac{w_i}{w_1 + w_2 + w_3} \qquad i = 0, 1, 2\,. \tag{2.87}$$

It can be readly seen, that the sum of these coordinates evaluates to one. Furthermore, barycentric coordinates are invariant under affine transformations, i.e., a linear transformation with a shift. A linear transformation can be encoded by

$$\hat{\boldsymbol{r}} = \mathsf{A} \cdot \boldsymbol{r}$$

with a second order tensor A. Applying a linear transformation together with a shift $\boldsymbol{s}$ to (2.86) and using (2.87) yields

$$\boldsymbol{s} + \hat{\boldsymbol{r}}_c = \lambda_0[\boldsymbol{s} + \hat{\boldsymbol{r}}_0] + \lambda_1[\boldsymbol{s} + \hat{\boldsymbol{r}}_1] + \lambda_2[\boldsymbol{s} + \hat{\boldsymbol{r}}_2]$$

with a constant vector $\boldsymbol{s}$. A simple algebraic manipulation of this intermediate result gives

$$\hat{\boldsymbol{r}}_c = \lambda_0 \hat{\boldsymbol{r}}_0 + \lambda_1 \hat{\boldsymbol{r}}_1 + \lambda_2 \hat{\boldsymbol{r}}_2 ,$$

which shows the invariance of the barycentric coordinates by comparison with (2.86).

First, linear interpolation on a reference triangle is considered. In the following, the space of all polynomials with $d_t \leq k$ and dimension N is called $\mathcal{P}^{k,N}(\Omega)$.

For $N = 2$ and using Cartesian coordinates ξ_1, ξ_2, it is straight-forward to define sets of basis functions according to the desired order of the polynomial as follows

$$\begin{aligned}
\mathcal{P}^{1,2} &: \{1,\ \xi_1,\ \xi_2\} , \\
\mathcal{P}^{2,2} &: \{1,\ \xi_1,\ \xi_2,\ \xi_1\xi_2,\ \xi_1^2,\ \xi_2^2\} , \\
\mathcal{P}^{3,2} &: \{1,\ \xi_1,\ \xi_2,\ \xi_1\xi_2,\ \xi_1^2,\ \xi_2^2,\ \xi_1^2\xi_2,\ \xi_1\xi_2^2,\ \xi_1^3,\ \xi_2^3\} ,
\end{aligned}$$

etc. The dimension of such vector spaces can be determined by

$$\dim(\mathcal{P}^{k,N}) = \frac{1}{N!} \prod_{i=1}^{N} (k+i) .$$

The barycentric coordinates for a unit triangle according to Fig. 2.5 (right) are given in terms of Cartesian coordinates by

$$\lambda_0 = 1 - \xi_1 - \xi_2 , \qquad \lambda_1 = \xi_1 , \qquad \lambda_2 = \xi_2 . \tag{2.88}$$

The unit triangle as well as linear, quadratic and cubic triangular Lagrange elements together with corresponding interpolation nodes are shown in Fig. 2.5 (right). The basis functions for linear and quadratic triangular Lagrange elements are given in Table 2.4 in terms of barycentric coordinates. In addition, the corresponding Cartesian coordinates for the unit triangle are given. The corresponding graphs are shown in Figs. 2.8 and 2.9, respectively.

2.3.5 *Remarks on Integration Using Reference Domains*

Reference domain $\Omega_\square$ and individual domains Ω_e are related via mappings χ_e. These mappings are in general non linear. Affine mappings, i.e., linear mappings with offset, are special cases.

Computation of stiffness matrices requires integration over the considered domain which is more conveniently carried out using the reference domain. The procedure is first illustrated for quadratic quadrilateral elements as shown in Fig. 2.10. In this case, the mapping χ_e from the reference domain $\Omega_\square$ to Ω_e

$$\begin{aligned}
\chi_e &: \Omega_\square \to \Omega_e \\
&(\xi_1, \xi_2) \mapsto \big(\chi_{e_1}(\xi_1, \xi_2),\ \chi_{e_2}(\xi_1, \xi_2)\big)
\end{aligned} \tag{2.89}$$

Table 2.4 Node coordinates and basis functions in terms of barycentric coordinates up to interpolation order two together with the corresponding Cartesian coordinates for the unit triangle (see, Fig: 2.5 (right)). For a graphical representation of the basis functions see Figs. 2.8 and 2.9

Order	Node i	Cartesian coordinates (ξ_1, ξ_2)	Barycentric coordinates $(\lambda_0, \lambda_1, \lambda_2)$	Basis function i
1	1	$(0, 0)$	$(1, 0, 0)$	$N_1 = \lambda_0$
	2	$(1, 0)$	$(0, 1, 0)$	$N_2 = \lambda_1$
	3	$(0, 1)$	$(0, 0, 1)$	$N_3 = \lambda_2$
2	1	$(0, 0)$	$(1, 0, 0)$	$N_1 = \lambda_0[2\lambda_0 - 1]$
	2	$(1, 0)$	$(0, 1, 0)$	$N_2 = \lambda_1[2\lambda_1 - 1]$
	3	$(0, 1)$	$(0, 0, 1)$	$N_3 = \lambda_2[2\lambda_2 - 1]$
	4	$(\frac{1}{2}, 0)$	$(\frac{1}{2}, \frac{1}{2}, 0)$	$N_4 = 4\lambda_0\lambda_1$
	5	$(0, \frac{1}{2})$	$(\frac{1}{2}, 0, \frac{1}{2})$	$N_5 = 4\lambda_0\lambda_2$
	6	$(\frac{1}{2}, \frac{1}{2})$	$(0, \frac{1}{2}, \frac{1}{2})$	$N_6 = 4\lambda_1\lambda_2$

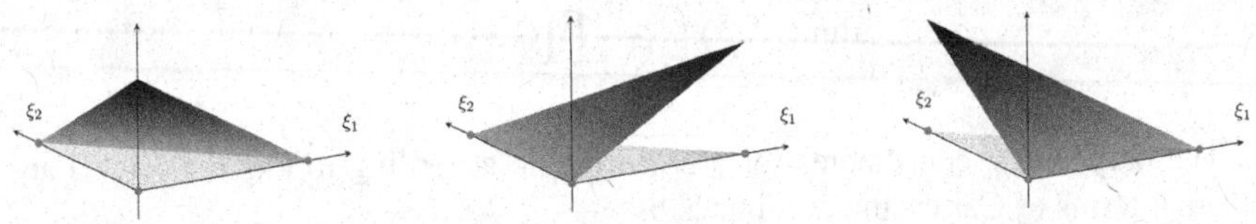

Fig. 2.8 Basis functions for the linear triangular element

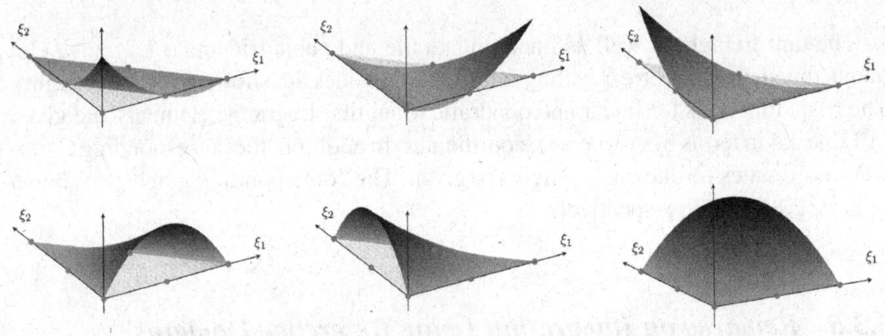

Fig. 2.9 Basis functions for the quadratic triangular element

reads the coordinates of a point which belongs to $\Omega_\Box$ and returns the coordinates of the corresponding point in Ω_e. Its linearisation is a linear mapping encoded by a second order tensor F_e, known as the Jacobian of χ_e, see Sect. B.11.

The Riemann integral of a function $\overline{f}(x_1, x_2)$ with respect to Ω_e requires first of all a partitioning of Ω_e into sub domains $\Delta\Omega_e^{(k)}$ of size $\mu\left(\Delta\Omega_e^{(k)}\right)$. The specific meaning of μ, hence size, depends whether integration is performed in one, two or three dimensions. For two dimensional domains, size refers to the domain's area. If the Riemann sum

$$S = \sum_{k=1}^{M} \overline{f}\left(x_1^{(k)}, x_2^{(k)}\right) \mu\left(\Delta\Omega_e^{(k)}\right)$$

with $(x_1^{(k)}, x_2^{(k)}) \in \Omega_e^{(k)}$ converges for $M \to \infty$, the result is called the integral of $f(x_1, x_2)$ over Ω_e and denoted by

$$\lim_{M\to\infty} S = \int_{\Omega_e} \overline{f}(x_1, x_2)\, \mathrm{d}^2 x\,,$$

provided that convergence does not depend on a particular choice regarding the points $\left(x_1^{(k)}, x_2^{(k)}\right)$. See, Appendix B, more specifically Sect. B.8, for details.

For a partitioning of $\Omega_\square$ into equally sized sub domains, it can be shown, that

$$\mu\left(\Delta\Omega_e^{(k)}\right) = |\det(\mathsf{F}_e)|\mu\left(\Delta\Omega_\square\right)\,, \tag{2.90}$$

from which

$$\int_{\Omega_e} \overline{f}(x_1, x_2)\, \mathrm{d}^2 x = \int_{\Omega_\square} f(\xi_1, \xi_2)\, |\det(\mathsf{F}_e)|\, \mathrm{d}^2\xi = \int_{\xi_1=-1}^{1} \int_{\xi_2=-1}^{1} f(\xi_1, \xi_2)\, |\det(\mathsf{F}_e)|\, \mathrm{d}\xi_1\, \mathrm{d}\xi_2$$

with $f = \overline{f} \circ \chi_e$ follows if $\Omega_\square$ is a unit square according to Fig. 2.5. For $\Omega_\square$ being the unit triangle shown in Fig. 2.5,

$$\int_{\Omega_e} \overline{f}(x_1, x_2)\, \mathrm{d}^2 x = \int_{\Omega_\square} f(\xi_1, \xi_2)\, |\det(\mathsf{F}_e)|\, \mathrm{d}^2\xi = \int_{\xi_1=0}^{1} \int_{\xi_2=0}^{1-\xi_1} f(\xi_1, \xi_2)\, |\det(\mathsf{F}_e)|\, \mathrm{d}\xi_2\, \mathrm{d}\xi_1$$

applies. Please note, that F_e cannot be the same for the unit square and the unit triangle, respectively.

The well-known relation (2.90) can be derived in a variety of ways. Here, the representation of area elements in terms of bivectors is employed. For details regarding exterior product and bivectors, see, Definition A.29. According to Fig. 2.10, $\Delta\Omega_e^{(k)}$ can be represented by

$$\Delta\Omega_e^{(k)} = \boldsymbol{A} \wedge \boldsymbol{B}\,.$$

The representation of the corresponding area element $\Delta\Omega_\square$ reads

$$\Delta\Omega_\square = \boldsymbol{a} \wedge \boldsymbol{b} = \Delta\xi_1 \Delta\xi_2\, \boldsymbol{e}_1 \wedge \boldsymbol{e}_2\,.$$

The areas of $\Delta\Omega_\square$ and $\Delta\Omega_e$ can be calculated by means of the bivector norm based on the corresponding inner product. According to (A.6), the area of $\Delta\Omega_\square$ is given

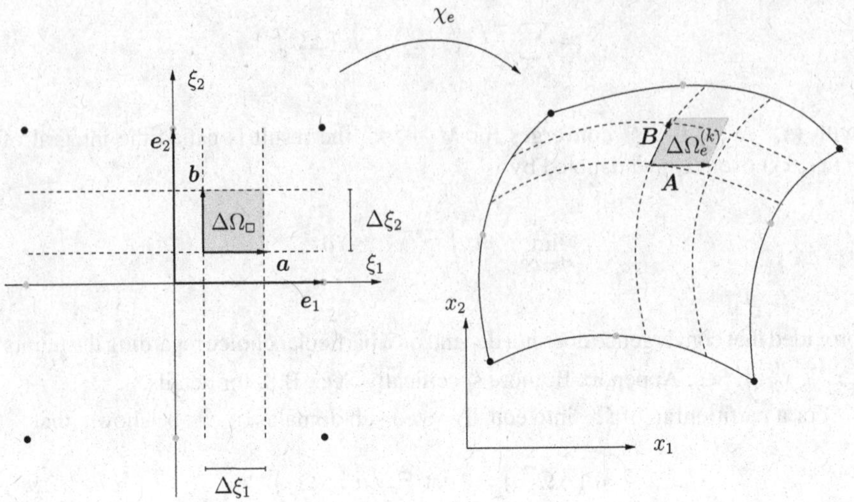

Fig. 2.10 Change of areas under transformations illustrated for some domain Ω_e (right) and a quadrilateral reference domain $\Omega_\square$ (left) with quadratic shape functions

by

$$\mu(\Delta\Omega_\square) = ||\boldsymbol{a} \wedge \boldsymbol{b}|| = |\Delta\xi_1 \Delta\xi_2| \, .$$

The vectors $\boldsymbol{A}$, $\boldsymbol{B}$ are related with $\boldsymbol{a}$ and $\boldsymbol{b}$ via the linear mapping encoded by F_e evaluated at the lower left corner of the $\Delta\Omega_\square$ defined by $\boldsymbol{a}$ and $\boldsymbol{b}$, i.e.,

$$\boldsymbol{A} = \mathsf{F}_e \cdot \boldsymbol{a} \qquad\qquad \boldsymbol{B} = \mathsf{F}_e \cdot \boldsymbol{b}$$

Therefore.

$$\begin{aligned}\boldsymbol{A} \wedge \boldsymbol{B} &= \left[\mathsf{F}_e \cdot \boldsymbol{a}\right] \wedge \left[\mathsf{F}_e \cdot \boldsymbol{b}\right] = \Delta\xi_1 \Delta\xi_2 \left[\mathsf{F}_e \cdot \boldsymbol{e}_1\right] \wedge \left[\mathsf{F}_e \cdot \boldsymbol{e}_2\right] = \Delta\xi_1 \Delta\xi_2 \, F_{1j} \, F_{2p} \boldsymbol{e}_j \wedge \boldsymbol{e}_p \\ &= [F_{11} F_{22} - F_{12} F_{21}] \Delta\xi_1 \Delta\xi_2 \, \boldsymbol{e}_1 \wedge \boldsymbol{e}_2\end{aligned}$$

holds and (2.90) follows immediately from the fact, that

$$\mu(\Delta\Omega_e) = ||\boldsymbol{A} \wedge \boldsymbol{B}|| = |\Delta\xi_1 \Delta\xi_2| \, |F_{11} F_{22} - F_{12} F_{21}| = |\Delta\xi_1 \Delta\xi_2| \, |\det(\mathsf{F}_e)| \, .$$

Please note, that $||\boldsymbol{A} \wedge \boldsymbol{B}||$ refers to the norm defined by (A.6). All remaining vertical lines in the preceding formula indicate absolute values.

The scheme extends in a straight forward manner for volume integration representing volume elements by trivectors.

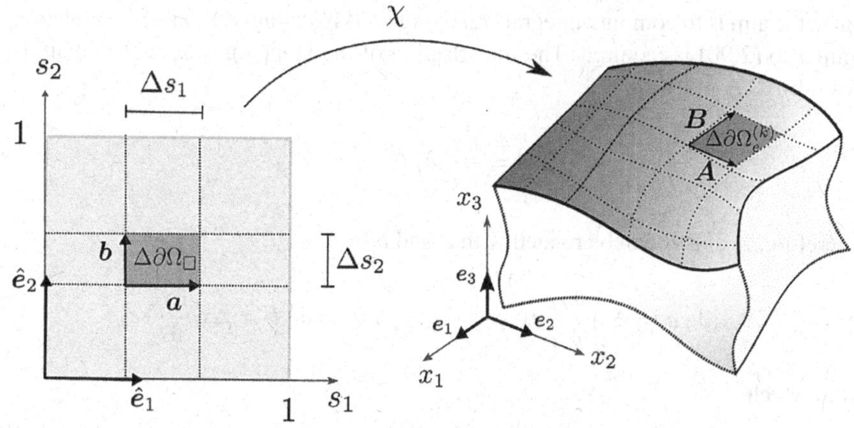

Fig. 2.11 Illustration of surface integration for a cuboid with nonlinear shape functions

Computation of load vectors requires integration with respect to the boundaries of domains Ω_e. These boundaries can either be surfaces or curves, depending whether $\Omega_e \subset \mathbb{R}^3$ or $\Omega_e \subset \mathbb{R}^2$, respectively.

The case $\Omega_e \subset \mathbb{R}^3$ is illustrated for a cuboid with nonlinear shape functions in Fig. 2.11. Surface integration is required if one or more of the six faces $\partial\Omega_e^{(i)}$ with $i = 1, \ldots 6$ of the cuboid belong to that part of the boundary of Ω with non-homogeneous natural boundary conditions.

A domain $\partial\Omega_\square \subset \mathbb{R}^2$ is considered which is related to a face $\partial\Omega_e^{(i)}$ of a cuboid by the mapping χ

$$\begin{aligned}\chi : & \partial\Omega_\square \to \partial\Omega_e^{(i)} \\ & (s_1, s_2) \mapsto (\chi_1 = x_1(s_1, s_2), \chi_2 = x_2(s_1, s_2), \chi_3 = x_3(s_1, s_2))\end{aligned}$$

which assigns coordinates x_1, x_2 and x_3 to points of $\partial\Omega_\square$ indicated by coordinates s_1 and s_2.

Furthermore, $\partial\Omega_\square$ is partitioned using rectangular area elements of equal size $\mu(\Delta\partial\Omega_\square)$. According to Fig. 2.11, these area elements can be represented by a bivector $\boldsymbol{a} \wedge \boldsymbol{b}$. Hence,

$$\mu(\Delta\partial\Omega_\square) = ||\boldsymbol{a} \wedge \boldsymbol{b}|| = |\Delta s_1 \Delta s_2| \, .$$

Since the aim is to compute integrals referring to $\partial\Omega_e^{(i)}$ using $\partial\Omega_\square$ instead, a relation similar to (2.90) is required. The linearisation of the mapping, i.e., its Jacobian, is given by

$$\mathsf{J} = \frac{\partial \chi_k}{\partial u_j} \hat{\boldsymbol{e}}_j \otimes \boldsymbol{e}_k \,.$$

Therefore, $\boldsymbol{A}$ and $\boldsymbol{B}$ can be related with $\boldsymbol{a}$ and $\boldsymbol{b}$ by

$$\boldsymbol{A} = \mathsf{J} \cdot \boldsymbol{a} = \Delta s_1 \frac{\partial \chi_k}{\partial s_1} \boldsymbol{e}_k \,, \qquad \boldsymbol{B} = \mathsf{J} \cdot \boldsymbol{b} = \Delta s_2 \frac{\partial \chi_m}{\partial s_2} \boldsymbol{e}_m$$

from which

$$\mu(\Delta\partial\Omega_e) = ||\boldsymbol{A} \wedge \boldsymbol{B}|| = \mu(\Delta\partial\Omega_\square) \left|\left| \frac{\partial \chi_k}{\partial s_1} \frac{\partial \chi_m}{\partial s_2} \boldsymbol{e}_k \wedge \boldsymbol{e}_m \right|\right| = \mu(\Delta\partial\Omega_\square)\, \beta$$

follows. Evaluating β explicitly yields

$$\beta = \sqrt{\beta_{12}^2 + \beta_{13}^2 + \beta_{23}^2} \tag{2.91}$$

with

$$\beta_{12} = \frac{\partial \chi_1}{\partial s_2} \frac{\partial \chi_2}{\partial s_1} - \frac{\partial \chi_2}{\partial s_1} \frac{\partial \chi_1}{\partial s_2} \,,$$
$$\beta_{13} = \frac{\partial \chi_1}{\partial s_2} \frac{\partial \chi_3}{\partial s_1} - \frac{\partial \chi_3}{\partial s_1} \frac{\partial \chi_1}{\partial s_2} \,,$$
$$\beta_{23} = \frac{\partial \chi_2}{\partial s_2} \frac{\partial \chi_3}{\partial s_1} - \frac{\partial \chi_3}{\partial s_1} \frac{\partial \chi_2}{\partial s_2} \,.$$

Therefore, if $\partial\Omega_\square$ is a square $(0, 1) \times (0, 1)$ with coordinates s_1 and s_2 as illustrated in Fig. 2.11, integration of some function $\overline{g}$ defined on a face $\partial\Omega_e^{(i)}$ of a three-dimensional finite element can be performed as follows

$$\int\limits_{\partial\Omega_e^{(i)}} \overline{g}\, \mathrm{d}S = \int\limits_{\partial\Omega_\square} g\, \beta\, \mathrm{d}^2 s = \int\limits_{s_1=0}^{1} \int\limits_{s_2=0}^{1} g\, \beta\, \mathrm{d}s_1\, \mathrm{d}s_2$$

with $g = \overline{g} \circ \chi$ and β given by (2.91).

2.4 Linear Elastostatics

2.4.1 Preliminary Remarks

Linear elastostatics forms part of the Theory of Elasticity which in turn is a sub-discipline of Continuum Mechanics. The latter intends to model motion and deformation of objects in space employing a continuum approach not only for space and time but as well for the objects in question. In its simplest version, Continuum Mechanics uses Euclidean geometry for modelling space, whereas time is modelled as a parameter, i.e., a real number. An object is modelled as a set of points with physical properties, such as density, temperature, velocity, etc., occupying at time t some connected region in space which is called the configuration of the object at a given time t.

As the name indicates, linear elastostatics does not really account for the dynamics of the problem but it only considers the change from some configuration in static equilibrium to another equilibrium configuration, caused, for instance, by applying displacements or surface tractions. Only small changes are permitted. Hence, only small displacements and small deformations are taken into account. Therefore, linear elastostatics does not distinguish between initial and final configuration in terms of global balance equations. In addition, deformations, i.e., changes in volume and shape, are quantified in terms of a strain tensor which depends only linearly on the displacement gradients.

Mechanical strains and stresses form a cause-effect couple for which Linear Elastostatics postulates a linear relation. More specifically, mechanical strain tensor and stress tensor are related by a fourth order elasticity tensor which maps mechanical strains to corresponding stresses.

The fact, that primary unknowns in Continuum Mechanics and, therefore, also in Linear Elastostatics, are in general vector fields increases the complexity compared with the boundary value problems discussed so far. Since the matter requires at least a working knowledge regarding tensor calculus, less advanced readers are recommended to examine at least the corresponding information provided in Appendices A and B in parallel while proceeding further.

2.4.2 Strong Form

Eventually, linear elastostatics is defined by the following system of equations

$$\nabla \cdot \sigma + \boldsymbol{f} = 0 \tag{2.92}$$

$$\varepsilon = \frac{1}{2}\left[\nabla \boldsymbol{u} + [\nabla \boldsymbol{u}]^{\mathrm{T}}\right] \tag{2.93}$$

$$\sigma = \mathsf{C} : \varepsilon \tag{2.94}$$

for the unknown displacement field $\boldsymbol{u}$ in a domain $\Omega \subset \mathrm{R}^3$, where σ, ε, C and $\boldsymbol{f}$ denote stress tensor, strain tensor, elasticity tensor, and volume force vector, respectively. The corresponding Dirichlet and Neumann boundary conditions read

$$\boldsymbol{u} = \overline{\boldsymbol{u}} \quad \text{at } \partial\Omega_u \tag{2.95}$$

$$\boldsymbol{t} = \overline{\boldsymbol{t}} \quad \text{at } \partial\Omega_t \tag{2.96}$$

in terms of displacement vector $\boldsymbol{u}$ and surface traction vector $\boldsymbol{t}$ and a bar indicates prescribed values for the corresponding variable. In classical continuum mechanics, surface tractions depend linearly on the unit normal vector $\boldsymbol{n}$

$$\boldsymbol{t} = \boldsymbol{\sigma} \cdot \boldsymbol{n}$$

which is known as Cauchy stress principle. The latter is an essential assumption required to derive (2.92) from the global force balance.

Furthermore, stress and strain tensor are symmetric. Therefore, it follows directly from (2.94), that the elasticity tensor C

$$\mathsf{C} = C_{ijkl}\boldsymbol{e}_i \otimes \boldsymbol{e}_j \otimes \boldsymbol{e}_k \otimes \boldsymbol{e}_l$$

must possess the following symmetries

$$C_{ijkl} = C_{jikl} = C_{ijlk} = C_{jilk} \,.$$

In addition, it can be shown, that

$$C_{ijkl} = C_{klij}$$

holds as well by inspecting the relations between stress, strain and strain energy density.

For a Cartesian coordinate system, taking into account all symmetries mentioned above, (2.92)–(2.94) can be written in index notation as follows

$$\sigma_{ij,j} + f_i = 0 \tag{2.97}$$

$$\varepsilon_{ij} = \frac{1}{2}[u_{i,j} + u_{j,i}] \tag{2.98}$$

$$\sigma_{ij} = C_{ijkl}\varepsilon_{kl} \tag{2.99}$$

whereas the boundary conditions read

$$u_i = \overline{u}_i \quad \text{at } \partial\Omega_u \tag{2.100}$$

$$t_i = \overline{t}_i \quad \text{at } \partial\Omega_t \tag{2.101}$$

together with Cauchy's stress principle

$$t_i = \sigma_{ij} n_j \,. \tag{2.102}$$

For an isotropic material, the constitutive relations are given by

$$\sigma_{ij} = \lambda \varepsilon_{kk} \delta_{ij} + 2\mu \varepsilon_{ij}$$

where λ and μ are the so-called Lamé constants.

Before deriving variational and weak form of the boundary value problem, two simplifications, known as plane stress and plane strain elastotatics are discussed.

2.4.3 *Strong Forms for Plane Stress and Plane Strain Problems*

If certain requirements are fulfilled in terms of geometry and loading conditions, two dimensional models are adequate for capturing reality with sufficient precision. The most common cases are known as plane stress and plane strain and examples are shown in Fig. 2.12.

The requirements for a plane stress model are outlined using the example on the left in Fig. 2.12 as an illustration. The following conditions have to be met:

1. The considered body should possess two traction free faces which are parallel to a symmetry plane Ω.
2. The distance between these faces should be significantly smaller than all other overall dimensions.
3. Boundary conditions must not vary in the direction perpendicular to Ω.

In terms of the coordinate system shown in Fig. 2.12, it follows from condition one and Cauchy's stress principle (2.102), that all stresses σ_{i3} must vanish at the top and the bottom surface, i.e.,

$$\sigma_{i3} = 0 \quad \text{for} \quad x_3 = \pm\frac{d}{2} \,.$$

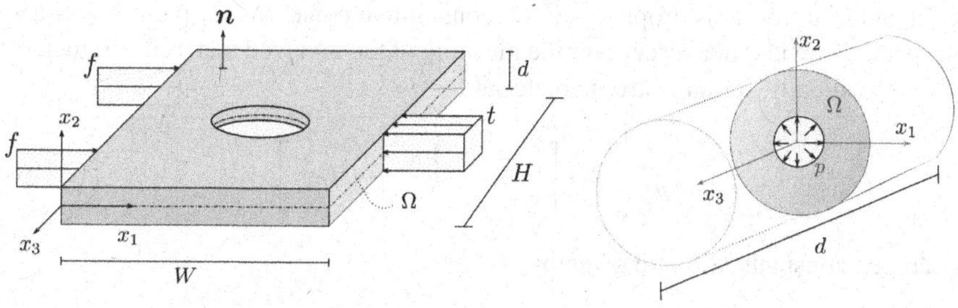

Fig. 2.12 Typical examples for simplifications in terms of plane stress (left) or plane strain (right)

Inside the body, there can still be stresses σ_{i3}. However, if the second condition is met as well, it is reasonable to assume, that $\sigma_{i3} = 0$ holds for the whole body. This in turn means that $\sigma_{i3} = 0$ due to the symmetry of the stress tensor. Together with the third condition, it follows eventually, that the solution does not depend on x_3 but only on x_1 and x_2. Therefore, a model which considers only the domain $\Omega \subset \mathbb{R}^2$ is sufficient.

Regarding plane strain, the reasoning is much simpler. The only conditions which have to be met are the following.

1. The displacement in one direction has to be zero.
2. Boundary conditions must not depend on this direction.

A standard example for plane strain is the thick-walled cylinder under internal pressure and fixed in axial direction at both ends, as illustrated on the right of Fig. 2.12. Evaluating the kinematic relations (2.98) for the coordinate system used here and $u_3 = 0$ yields

$$\varepsilon_{i3} = 0\,,$$

which ensures together with condition one, that, again, a model which considers only the domain $\Omega \subset \mathbb{R}^2$ is sufficient.

The constitutive relations for plane strain and plane stress differ, whereas equilibrium conditions and kinematics are only affected insofar as $\Omega \subset \mathbb{R}^2$ and not $\Omega \subset \mathbb{R}^3$. Hence the set of equations to be fulfilled within $\Omega \subset \mathbb{R}^2$ reads for plane stress and plane strain

$$\begin{aligned}\sigma_{ij,j} + f_i &= 0\\ \varepsilon_{ij} &= \frac{1}{2}[u_{i,j} + u_{j,i}]\\ \sigma_{ij} &= C^{\mathrm{E/S}}_{ijkl}\varepsilon_{kl}\end{aligned} \tag{2.103}$$

for $i, j, k, l = 1, 2$, where 'E/S' indicates, that the elasticity tensor has to be chosen according to the considered case, i.e., plane strain (E) or plane stress (S).

The boundary conditions have the same structure given by (2.100) and (2.101) but $\partial\Omega$ is now a curve in $\mathbb{R}^2$ and indexes take only values up to two.

In addition, for an isotropic material, constitutive equations for plane stress and plane strain are identical except for the meaning of the involved material parameters. More specifically, for an isotropic material

$$C^{\mathrm{E/S}}_{ijkl} = 2G\left[\delta_{ik}\delta_{jl} - \frac{1}{2}\frac{\kappa - 3}{\kappa - 1}\delta_{kl}\delta_{ij}\right] \tag{2.104}$$

where the constants G, κ are given by

$$G = \frac{E}{2[1+\nu]}\,, \quad \kappa = \begin{cases} 3-4\nu & \text{plane strain} \\ \frac{3-4\nu}{1+\nu} & \text{plane stress} \end{cases}$$

with Young's modulus E, Poisson's ratio ν, and shear modulus G. For details see, for instance, [1]. In addition, the information regarding the third direction is given by

$$\varepsilon_{33} = -\frac{\nu}{E}\,[\sigma_{11} + \sigma_{22}]$$

for plane stress and

$$\sigma_{33} = -\frac{\nu E}{[1+\nu][1-2\nu]}\,[\varepsilon_{11} + \varepsilon_{22}]$$

for plane strain. It should be noted that although $\varepsilon_{33} \neq 0$ for plane stress and $\sigma_{33} \neq 0$ for plane strain, the corresponding terms in the product $\sigma_{ij}\varepsilon_{ij}$ cancel out because $\sigma_{33} = 0$ and $\varepsilon_{33} = 0$ for plane stress and plane strain, respectively.

2.4.4 Variational and Weak Forms

The primary unknown in Linear Elastostatics is the displacement field $\boldsymbol{u}$, which is a vector field also called a vector-valued function. To derive the variational form, we consider the integral of the product of (2.97) with the coordinate functions of a test vector field $\boldsymbol{v}$ with respect to the domain $\Omega \subset \mathbb{R}^3$

$$\mathcal{W} = \int\limits_{\Omega} \left[\sigma_{ij,j} + f_i\right] v_i \, \mathrm{d}^3x = 0\,. \tag{2.105}$$

From

$$\int\limits_{\Omega} \sigma_{ij,j} v_i \, \mathrm{d}^3x = \int\limits_{\Omega} (\sigma_{ij} v_i)_{,j} \, \mathrm{d}^3x - \int\limits_{\Omega} \sigma_{ij} v_{i,j} \, \mathrm{d}^3x$$

follows by applying Gauss theorem (see, (B.18)) to the first term of the right hand side

$$\int\limits_{\Omega} \sigma_{ij,j} v_i \, \mathrm{d}^3x = \int\limits_{\partial\Omega} \sigma_{ij} v_i n_j \, \mathrm{d}S - \int\limits_{\Omega} \sigma_{ij} v_{i,j} \, \mathrm{d}^3x\,.$$

Taking into account Cauchy's stress principle (2.102), yields

$$\int_{\Omega} \sigma_{ij,j} v_i \, \mathrm{d}^3x = \int_{\partial\Omega} t_i v_i \, \mathrm{d}S - \int_{\Omega} \sigma_{ij} v_{i,j} \, \mathrm{d}^3x \,. \tag{2.106}$$

By means of (2.106), (2.105) can be written as

$$\mathcal{W} = \int_{\Omega} \sigma_{ij} v_{i,j} \, \mathrm{d}^3x - \int_{\Omega} f_i v_i \, \mathrm{d}^3x - \int_{\partial\Omega_u} t_i v_i \, \mathrm{d}S - \int_{\partial\Omega_t} t_i v_i \, \mathrm{d}S = 0 \,, \tag{2.107}$$

because $\partial\Omega = \partial\Omega_t \bigcup \Omega_u$. Inspection of equation (2.107) reveals, that the boundary condition (2.101) can be plugged directly into the most right hand side term. Due to the symmetry of the stress tensor

$$\sigma_{ij} v_{i,j} = \sigma_{ij} \frac{1}{2} [v_{i,j} + v_{j,i}]$$

holds which motivates the definition of a strain operator

$$\eta_{ij}(\boldsymbol{w}) := \frac{1}{2} [w_{i,j} + w_{j,i}] \,. \tag{2.108}$$

Finally, considering only test function vectors $\boldsymbol{v}$ which vanish at the boundary $\partial\Omega_u$ yields eventually the variational form of the boundary value problem

$$\mathcal{W} = \int_{\Omega} C_{ijkl} \eta_{i,j}(\boldsymbol{u}) \, \eta_{k,l}(\boldsymbol{v}) \, \mathrm{d}^3x - \int_{\Omega} f_i v_i \, \mathrm{d}^3x - \int_{\partial\Omega_t} \bar{t}_i v_i \, \mathrm{d}S = 0 \,, \tag{2.109}$$

$$u_i = \bar{u}_i \quad \text{on } \partial\Omega_u \tag{2.110}$$

and the weak form is obtained from it by interpreting partial derivatives in the weak sense.

A suitable test function space W consists of all weakly partially differentiable vector fields $\boldsymbol{f} : \Omega \to \mathbb{R}^3$. Limiting ourselves to Cartesian vectors, i.e. $\mathbb{R}^3$ with standard basis,

$$W(\Omega) = \{\phi \mid \phi \in H^1(\Omega) \text{ and } \phi = 0 \text{ on } \partial\Omega_u\} \,. \tag{2.111}$$

is defined. Trial functions, on the other hand, must fulfil the Dirichlet boundary conditions (2.95), which are not necessarily homogeneous. Therefore, the spaces

$$V_i(\Omega) = \{g \mid g \in H^1(\Omega) \text{ and } g = \bar{u}_i \text{ on } \partial\Omega_u\} \,, \tag{2.112}$$

with $i = 1, 2, 3$, are defined as trial function spaces. In view of the Lax-Milgram lemma, the spaces $V = V_1 \times V_2 \times V_3$ and $W^3 = W \times W \times W$ are required. The task of finding weak solutions for Linear Elastostatics can be stated as follows.

Task 2.6 Find $\boldsymbol{u}$ with $u_i \in V_i(\Omega)$ according to (2.112) such that

$$\mathcal{W} = \int_{\Omega} C_{ijkl}\eta_{ij}(\boldsymbol{u})\,\eta_{kl}(\boldsymbol{v})\,\mathrm{d}^3x - \int_{\Omega} f_i v_i\,\mathrm{d}^3x - \int_{\partial\Omega_t} \bar{t}_i v_i\,\mathrm{d}S = 0$$

for arbitrary $v_i \in W$ defined by (2.111) and given $f_i, \bar{t}_i, C_{ijkl} \in L^2(\Omega)$.

If homogeneous Dirichlet boundary conditions are prescribed, i.e., $\bar{\boldsymbol{u}} = \boldsymbol{0}$ in (2.95), the spaces V_i and W coincide and the Lax-Milgram lemma applies. Otherwise, analogously to Sect. 2.1.8, Task 2.6 can be reformulated for homogeneous Dirichlet boundary conditions by defining $U_i = u_i - w_i$ with $w_i \in H^1(\Omega)$. Since $\Omega \in \mathbb{R}^3$, assigning boundary conditions involves the so-called trace theorem, see Sect. C.9. The reformulation of Task 2.6 is omitted here, because it is equivalent to a reorganisation of the global system of equations in the FEM solution procedure as discussed in Sect. 2.1.8.

Analogously, the variational forms for plane strain and plane stress can be derived. Apart from the fact, that in this case indexes take only values up to two, volume integration reduces to integrating over $\Omega \subset \mathbb{R}^2$ and multiplying the result with a length or thickness d. A similar procedure applies for the boundary integral. The meanings of Ω and d depend on whether plane stress or plane strain is considered, see Fig. 2.12. Dividing the result by d yields a formulation per unit thickness as follows

$$\mathcal{W} = \int_{\Omega} C^{\mathrm{E/S}}_{ijkl}\eta_{kl}(\boldsymbol{u})\,\eta_{ij}(\boldsymbol{v})\,\mathrm{d}^2x - \int_{\Omega} f_i v_i\,\mathrm{d}^2x - \int_{\partial\Omega_t} \bar{t}_i v_i\,\mathrm{d}S = 0\,, \tag{2.113}$$

$$u_i = \bar{u}_i \quad \text{on}\ \partial\Omega_u \tag{2.114}$$

for $i, j, k, l = 1, 2$, which will be used in Sect. 2.4.5.

2.4.5 *Galerkin FEM for Plane Strain/Stress Problems*

Eventually, FEM solution schemes have to be implemented as computer codes. The efficiency of such computer codes depends, among other things, on the number of operations required to execute specific tasks and the amount of data involved. There are a number of symmetries in Linear Elastostatics, which should be taken into account by an efficient implementation. There are at least two methods, known as Voigt notation and Mandel notation, to achieve this goal. Here, Voigt notation is used due to its popularity.

Using Voigt notation, the essential information about stresses and strains are provided in terms of one-column matrices as follows.

$$\underline{\boldsymbol{\sigma}}^{\mathrm{T}} = \left[\sigma_{11}\ \sigma_{22}\ \sigma_{33}\ \sigma_{23}\ \sigma_{13}\ \sigma_{12}\right],$$
$$\underline{\boldsymbol{\varepsilon}}^{\mathrm{T}} = \left[\varepsilon_{11}\ \varepsilon_{22}\ \varepsilon_{33}\ 2\varepsilon_{23}\ 2\varepsilon_{13}\ 2\varepsilon_{12}\right].$$

Multiplying the shear strains by a factor two is necessary to ensure

$$\sigma_{ij}\varepsilon_{ij} = \underline{\boldsymbol{\varepsilon}}^{\mathrm{T}}\,\underline{\boldsymbol{\sigma}},$$

because a change in notation or storage must not alter the results of a computation.

For plane stress and plane strain problems, the necessary information reduces to

$$\underline{\boldsymbol{\sigma}}^{\mathrm{T}} = \left[\sigma_{11}\ \sigma_{22}\ \sigma_{12}\right],$$
$$\underline{\boldsymbol{\varepsilon}}^{\mathrm{T}} = \left[\varepsilon_{11}\ \varepsilon_{22}\ 2\varepsilon_{12}\right].$$

In addition, the essential information regarding the elasticity tensor $\mathsf{C}^{\mathrm{E/S}}$ can be given in matrix form. Therefore, the constitutive relations can be expressed by

$$\underline{\boldsymbol{\sigma}} = \underline{\underline{\boldsymbol{C}}}^{\mathrm{E/S}}\,\underline{\boldsymbol{\varepsilon}}$$

with

$$\underline{\underline{\boldsymbol{C}}}^{\mathrm{E/S}} = \begin{bmatrix} C^{\mathrm{E/S}}_{1111} & C^{\mathrm{E/S}}_{1122} & C^{\mathrm{E/S}}_{1112} \\ C^{\mathrm{E/S}}_{2211} & C^{\mathrm{E/S}}_{2222} & C^{\mathrm{E/S}}_{2212} \\ C^{\mathrm{E/S}}_{1211} & C^{\mathrm{E/S}}_{1222} & C^{\mathrm{E/S}}_{1212} \end{bmatrix}.$$

Assembling the coordinate functions of vector fields with respect to the standard basis in $\mathbb{R}^N$ as usual in one-column matrices, the weak form for plane strain/stress problems of Linear Elastostatics reads

$$\mathcal{W} = \int\limits_{\Omega} \underline{\boldsymbol{\eta}}^{\mathrm{T}}(\boldsymbol{v})\,\underline{\underline{\boldsymbol{C}}}^{\mathrm{E/S}}\underline{\boldsymbol{\eta}}(\boldsymbol{u})\,\mathrm{d}^2x - \int\limits_{\Omega} \underline{\boldsymbol{v}}^{\mathrm{T}}\,\underline{\boldsymbol{f}}\,\mathrm{d}^2x - \int\limits_{\partial\Omega_t} \underline{\boldsymbol{v}}^{\mathrm{T}}\,\underline{\bar{\boldsymbol{t}}}\,\mathrm{d}S = 0, \tag{2.115}$$

$$\underline{\boldsymbol{u}} = \underline{\bar{\boldsymbol{u}}} \text{ on } \partial\Omega_u \tag{2.116}$$

where Voigt notation is applied as well for the strain operator (2.108).

First, a discretisation is supposed by which Ω is divided into subdomains Ω_e such that

$$\Omega \approx \bigcup_{e=1}^{N_e} \Omega_e\,.$$

Integrals in (2.115) are evaluated in a piecewise manner, hence

$$\mathcal{W} = \sum_{e=1}^{N_e} \mathcal{W}_e \tag{2.117}$$

with

$$\mathcal{W}_e = \int\limits_{\Omega_e} \underline{\eta}^{\mathrm{T}}(\boldsymbol{v})\, \underline{\underline{C}}^{\mathrm{E/S}} \underline{\eta}(\boldsymbol{u})\, \mathrm{d}^2x - \int\limits_{\Omega_e} \underline{v}^{\mathrm{T}}\, \underline{f}\, \mathrm{d}^2x - \int\limits_{\partial\Omega_{e(t)}} \underline{v}^{\mathrm{T}}\, \underline{\bar{t}}\, \mathrm{d}S\,. \tag{2.118}$$

In the following, the FEM solution scheme for linear triangular elements is discussed. As in Sect. 2.2.5, a reference domain $\Omega_\square$ is used. Every Ω_e is related with the reference domain $\Omega_\square$ by a mapping χ_e, which is actually a coordinate transform given by

$$\chi_e : \Omega_\square \rightarrow \Omega_e \tag{2.119}$$

$$(\xi_1, \xi_2) \mapsto \left(\chi_{e_1} = \sum_{\alpha=1}^{3} N_\alpha(\xi_1, \xi_2)\, x_{1\alpha}\,,\ \chi_{e_2} = \sum_{\alpha=1}^{3} N_\alpha(\xi_1, \xi_2)\, x_{2\alpha} \right).$$

According to (2.88), the functions $N_\alpha(\xi_1, \xi_2)$ are given by

$$N_1 = 1 - \xi_1 - \xi_2\,, \qquad N_2 = \xi_1\,, \qquad N_3 = \xi_2\,. \tag{2.120}$$

For details, see Sect. 2.3.3. An example triangulation is shown in Fig. 2.13.

For trial displacements and test vector coordinate functions in $\Omega_\square$ we set

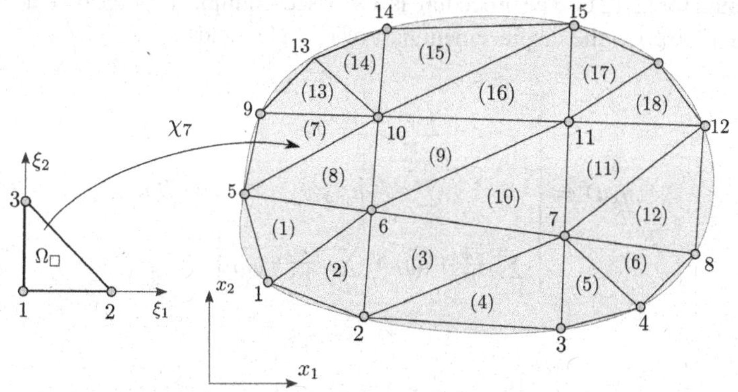

Fig. 2.13 Example triangulation of a two-dimensional domain Ω together with the reference domain $\Omega_\square$. Exemplary, a particular mapping χ_e is indicated

$$\underline{\boldsymbol{u}}_{\square} = \sum_{\alpha=1}^{4} N_{\alpha}(\xi_1, \xi_2)\, \underline{\hat{\boldsymbol{u}}}_{(\alpha)} \qquad \underline{\boldsymbol{v}}_{\square} = \sum_{\alpha=1}^{4} N_{\alpha}(\xi_1, \xi_2)\, \underline{\hat{\boldsymbol{v}}}_{(\alpha)} \tag{2.121}$$

where the $\underline{\hat{\boldsymbol{u}}}_{(\alpha)}$ refer to the displacements at the nodes with local numbers α. For example,

$$\underline{\hat{\boldsymbol{u}}}^{\mathrm{T}}_{(3)} = \begin{bmatrix} u_{1(3)} & u_{2(3)} \end{bmatrix}$$

with the displacements in ξ_1 and ξ_2 direction at the local node 3 indicated by $u_{1(3)}$ and $u_{2(3)}$, respectively. This scheme applies analogously for $\underline{\hat{\boldsymbol{v}}}_{(\alpha)}$. For the sake of a more efficient matrix notation,

$$\underline{\hat{\boldsymbol{u}}}^{\mathrm{T}}_{\square} = \begin{bmatrix} u_{1(1)} & u_{2(1)} & u_{1(2)} & u_{2(2)} & u_{1(3)} & u_{2(3)} \end{bmatrix}$$

and

$$\underline{\hat{\boldsymbol{v}}}^{\mathrm{T}}_{\square} = \begin{bmatrix} v_{1(1)} & v_{2(1)} & v_{1(2)} & v_{2(2)} & v_{1(3)} & v_{2(3)} \end{bmatrix}$$

are defined together with the matrix

$$\underline{\underline{\boldsymbol{N}}} = \begin{bmatrix} N_1 & 0 & N_2 & 0 & N_3 & 0 \\ 0 & N_1 & 0 & N_2 & 0 & N_3 \end{bmatrix}$$

to write (2.121) as

$$\underline{\boldsymbol{u}}_{\square} = \underline{\underline{\boldsymbol{N}}}\, \underline{\hat{\boldsymbol{u}}}_{\square}\,, \qquad \underline{\boldsymbol{v}}_{\square} = \underline{\underline{\boldsymbol{N}}}\, \underline{\hat{\boldsymbol{v}}}_{\square}\,. \tag{2.122}$$

The strain operators in (2.115) contain gradients of their arguments, which have to be evaluated for (2.121). The procedure is discussed exemplary for the displacements. Applying (2.108) to the displacement part of (2.121) yields

$$\underline{\boldsymbol{\eta}}(\boldsymbol{u}) = \begin{bmatrix} \sum\limits_{\alpha=1}^{3} \frac{\partial N_{\alpha}}{\partial x_1} u_{1(\alpha)} \\ \sum\limits_{\alpha=1}^{3} \frac{\partial N_{\alpha}}{\partial x_2} u_{2(\alpha)} \\ \sum\limits_{\alpha=1}^{3} \frac{\partial N_{\alpha}}{\partial x_2} u_{1(\alpha)} + \sum\limits_{\alpha=1}^{3} \frac{\partial N_{\alpha}}{\partial x_1} u_{2(\alpha)} \end{bmatrix} = \underline{\underline{\boldsymbol{B}}}\, \underline{\hat{\boldsymbol{u}}}_{\square}$$

with

$$\underline{\underline{\boldsymbol{B}}} = \begin{bmatrix} \underline{\underline{\boldsymbol{B}}}_1 & \underline{\underline{\boldsymbol{B}}}_2 & \underline{\underline{\boldsymbol{B}}}_3 \end{bmatrix}$$

and the sub-matrices

$$\underline{\underline{B}}_{\alpha} = \begin{bmatrix} \frac{\partial N_\alpha}{\partial x_1} & 0 \\ 0 & \frac{\partial N_\alpha}{\partial x_2} \\ \frac{\partial N_\alpha}{\partial x_2} & \frac{\partial N_\alpha}{\partial x_1} \end{bmatrix} .$$

The procedure applies analogously to $\underline{\eta}(\boldsymbol{v})$. Hence,

$$\underline{\eta}(\boldsymbol{u}) = \underline{\underline{B}}\,\underline{\hat{u}}_\square\,, \qquad \underline{\eta}(\boldsymbol{v}) = \underline{\underline{B}}\,\underline{\hat{v}}_\square\,. \tag{2.123}$$

By means of (2.122) and (2.123), the contribution of an element e to the weak form (2.117) can be written as follows

$$\mathcal{W}_{e_\square} = \underline{\hat{v}}_\square^{\mathrm{T}} \int_{\Omega_e} \underline{\underline{B}}^{\mathrm{T}}\, \underline{\underline{C}}^{\mathrm{E/S}}\, \underline{\underline{B}}\, \mathrm{d}^2x\, \underline{\hat{u}}_\square - \underline{\hat{v}}_\square^{\mathrm{T}} \int_{\Omega_e} \underline{\underline{N}}^{\mathrm{T}}\, \underline{f}\, \mathrm{d}^2x - \underline{\hat{v}}_\square^{\mathrm{T}} \int_{\partial\Omega_{e(t)}} \underline{\underline{N}}^{\mathrm{T}}\, \underline{\bar{t}}\, \mathrm{d}S\,,$$

or, more concisely as

$$\mathcal{W}_{e_\square} = \underline{\hat{v}}_\square^{\mathrm{T}} \left[\underline{\underline{K}}_e\, \underline{\hat{u}}_\square - \underline{r}_e - \underline{q}_e \right]$$

with element stiffness matrix $\underline{\underline{K}}_e$, volume force contribution $\underline{r}_e$, and prescribed surface traction contribution $\underline{q}_e$ given by

$$\underline{\underline{K}}_e = \int_{\Omega_e} \underline{\underline{B}}^{\mathrm{T}}\, \underline{\underline{C}}^{\mathrm{E/S}}\, \underline{\underline{B}}\, \mathrm{d}^2x \qquad \underline{r} = \int_{\Omega_e} \underline{\underline{N}}^{\mathrm{T}}\, \underline{f}\, \mathrm{d}^2x \qquad \underline{q} = \int_{\partial\Omega_{e(t)}} \underline{\underline{N}}^{\mathrm{T}}\, \underline{\bar{t}}\, \mathrm{d}S\,.$$

Again, the aim is to integrate with respect to the reference domain $\Omega_\square$. This is addressed in more detail in Sect. 2.3.5.

Finally, a discrete version of (2.117) can be written concisely as

$$\mathcal{W}_h = \underline{\hat{v}}^{\mathrm{T}} \left[\underline{\underline{K}}\,\underline{\hat{u}} - \underline{r} - \underline{q} \right] = 0 \tag{2.124}$$

with

$$\underline{\underline{K}} = \mathop{\mathsf{A}}_{e=1}^{N_e} \underline{\underline{K}}_e\,, \qquad \underline{r} = \mathop{\mathsf{A}}_{e=1}^{N_e} \underline{r}_e\,, \qquad \underline{q} = \mathop{\mathsf{A}}_{e=1}^{N_e} \underline{q}_e\,.$$

As before, the assembling operator A adds the element contributions to the global system, ensuring the correct correspondences.

Due to the arbitrariness of the test function values at the nodes, it follows from (2.124), that

$$\underline{\underline{K}}\,\underline{\hat{u}} = \underline{r} + \underline{q}\,. \tag{2.125}$$

After incorporating essential boundary conditions, the system of linear equations (2.125) can be solved to determine the unknown vector $\underline{\hat{u}}$. Afterwards, derived quantities such as stresses and strains can be computed by corresponding postprocessing routines.

2.5 Boundary Value Problems of Fourth Order in $\mathbb{R}$

2.5.1 *Strong Form*

The Euler-Bernoulli beam from strength of materials is the most widely used application of a fourth order bvp. Based on the usual three cornerstones of strength of materials, namely equilibrium, material law, and kinematics, the differential equation for the deflection of the beam

$$\left(E I w''\right)'' - q = 0 \tag{2.126}$$

is derived, which defines together with a complete set of boundary conditions the strong form of the bvp. For the Example 2.11, the boundary conditions read

$$w(0) = 0 \tag{2.127}$$
$$w'(0) = 0 \tag{2.128}$$
$$V(l) = -(E I w'')'(l) = F \tag{2.129}$$
$$M(l) = -E I w''(l) = M_l \tag{2.130}$$

with bending moment M, shear force V, Young's modulus E and area moment of inertia I. The product EI is known as beam stiffness and it can depend on spatial position. As long as EI does not depend on the deflection w, the boundary value problem is linear.

Example 2.11 We consider a beam of length l, fixed at the left and, exposed to a line load q. At the right end, a concentrated force and a moment are prescribed.

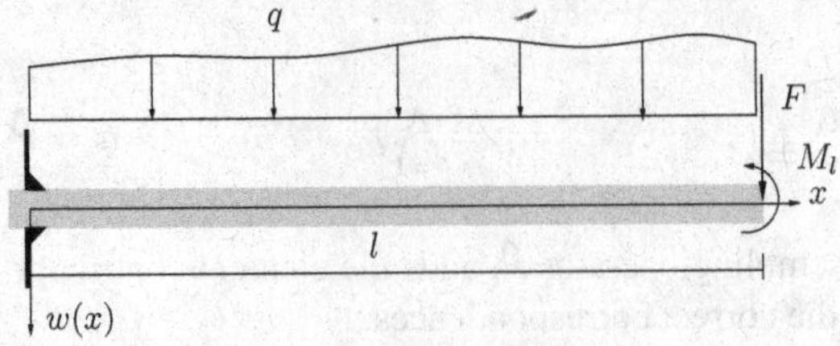

2.5.2 Variational and Weak Form

Analogously to the previous sections, our first goal is to derive the variational form of the boundary value problem. Therefore, the differential equation (2.126) is multiplied with a test function v and integrated over the domain for which the problem is defined

$$\int_0^l \left[\left(EIw''\right)'' - q \right] v \, \mathrm{d}x = 0 \, . \tag{2.131}$$

For clarity of presentation, we set $m = EIw''$ and perform integration by parts twice, i.e.,

$$\int_0^l m''v \, \mathrm{d}x = \underbrace{\int_0^l (m'v)' \, \mathrm{d}x}_{m'v|_0^l} - \int_0^l m'v' \, \mathrm{d}x = m'v|_0^l - \Big[\underbrace{\int_0^l (mv')' \, \mathrm{d}x}_{mv'|_0^l} - \int_0^l mv'' \, \mathrm{d}x \Big]$$

$$= \int_0^l mv'' \, \mathrm{d}x + m'v|_0^l - mv'|_0^l \, . \tag{2.132}$$

By means of (2.132), (2.131) eventually reads

$$\int_0^l EIw''v'' \, \mathrm{d}x - \int_0^l qv \, \mathrm{d}x + [\underbrace{(EIw'')'\big|_{x=l}}_{-V(l)} v(l) - \underbrace{(EIw'')'\big|_{x=0}}_{-V(0)} v(0)]$$

$$- [\underbrace{EIw''\big|_{x=l}}_{-M(l)} v'(l) - \underbrace{EIw''\big|_{x=l}}_{-M(0)} v'(0)] = 0 \, . \tag{2.133}$$

Similar to the rod under axial loading, there are boundary conditions which can be plugged into this equation directly, namely (2.129) and (2.130). According to the convention agreed upon before, this boundary conditions are called natural boundary conditions, whereas the remaining boundary conditions are again called essential boundary conditions. The remaining boundary terms vanish, if test functions are used which vanish everywhere, where essential boundary conditions are prescribed. With regard to Example 2.11, this means $v(0) = 0$ and $v'(0) = 0$. Therefore, the variational form for the Example 2.11 reads

$$\mathcal{W} = \int_0^l EIw''v''\,\mathrm{d}x - \int_0^l qv\,\mathrm{d}x - Fv(l) + M_l v'(l) = 0 \tag{2.134}$$

$$w(0) = 0$$
$$w'(0) = 0$$

and the weak form is obtained by interpreting the derivatives under the integral signs in a weak sense.

The space for trial and test functions for the weak form based on (2.134) is given by

$$V(\Omega) = \{f \mid f \in H^2(\Omega) \text{ and } f(0) = 0,\ f'(0) = 0\}, \tag{2.135}$$

where, by abuse of notation, the prime is interpreted as weak derivative. Please note that, since (2.134) contains second order derivatives, $V(\Omega)$ has to be defined based on $H^2(\Omega)$. The corresponding task of solving the weak is stated below.

Task 2.7 Find $w \in V(\Omega)$ according to (2.135) such that

$$\mathcal{W} = \int_0^l EIw''v''\,\mathrm{d}x - \int_0^l qv\,\mathrm{d}x - Fv(l) = 0$$

for arbitrary $v \in V(\Omega)$ and given $q, EI \in L^2(\Omega)$.

2.5.3 Galerkin FEM Using C^1 Continuity

Again, we are looking for an approximation in a finite dimensional subspace $V_h(\Omega)$ of $V(\Omega)$ defined by (2.135). The corresponding modification of Task 2.7 reads as follows.

Task 2.8 Find $w \in V_h(\Omega) \subset V(\Omega)$ according to (2.135) such that

$$\mathcal{W}_h = \int_0^l EIw_h''v_h''\,\mathrm{d}x - \int_0^l qv_h\,\mathrm{d}x - Fv_h(l) = 0$$

for arbitrary $v \in V_h(\Omega)$ and given $q, EI \in L^2(\Omega)$.

Our aim is again to solve approximately the weak form using piecewise defined trial and test functions. Piecewise linear functions can not be used here because their second weak derivative does not exist. As before, the use of piecewise defined functions implies the existence of a discretization, i.e., a division of the interval for which the problem is defined into subintervals. This, in turn, motivates

$$\mathcal{W}_h = \sum_{k=1}^{N} \mathcal{W}_e - F v_h(l) + M v_h'(l) \tag{2.136}$$

with

$$\mathcal{W}_e = \int_{x_e}^{x_{e+1}} EI w_h'' v_h'' \,\mathrm{d}x - \int_{x_e}^{x_{e+1}} q v_h \,\mathrm{d}x\,. \tag{2.137}$$

In the following, the reference domain concept, introduced already in Sect. 2.1.7, is used. Since domains are characterized by only two pieces of information, namely x_e and x_{e+1}, the mapping (2.34) and its inverse apply here as well. For the sake of clarity, a part of (2.34) is repeated

$$\chi_e : \xi \mapsto x = \frac{1}{2}(1-\xi)x_e + \frac{1}{2}(1+\xi)x_{e+1}\,.$$

In order to assure the required continuity, not only the values of the trial function at the left and right interval borders have to be taken into account but, as well the values of the first derivative of the trial functions at this locations. Hence, there are four informations the ansatz for the deflection within the reference domain must account for. Therefore, we set

$$w_\square(\xi) = C_0 + C_1\xi + C_2\xi^2 + C_3\xi^3$$

and in order to express the coefficients C_0, C_1, C_2, C_3 by the deflections and rotations at the left and right border, i.e., w_0, w_0', w_1, w_1' the following conditions are used

$$w_\square(\xi=-1) = w_0 \qquad w_\square(\xi=+1) = w_1$$
$$w_\square'(\xi=-1) = w_0' \qquad w_\square'(\xi=+1) = w_1'$$

where again, chain rule, i.e.,

$$w_\square'(\xi) = \frac{\mathrm{d}}{\mathrm{d}\xi}\tilde{w}_\square(\xi)\,\frac{\mathrm{d}\xi}{\mathrm{d}x} = \frac{\mathrm{d}}{\mathrm{d}\xi}\tilde{w}_\square(\xi)\, j_e^{-1}$$

applies, because of the coordinate transform. Solving the linear system of equations and grouping terms accordingly eventually yields

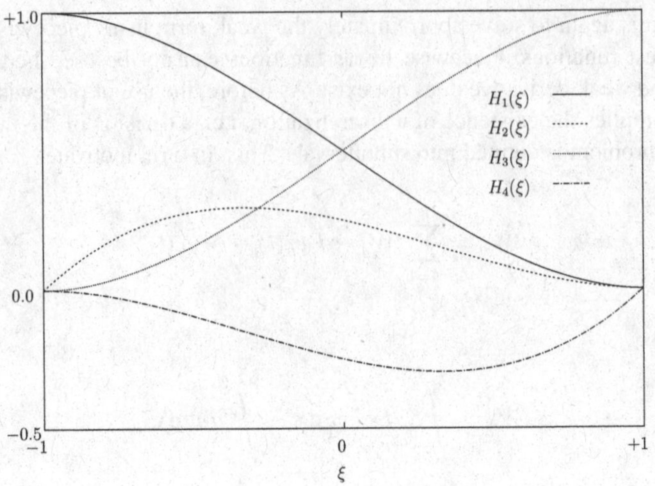

Fig. 2.14 Hermite-polynomials in the reference domain $\Omega_\square$

$$w_\square(\xi) = H_1(\xi)w_0 + H_2(\xi)\frac{L_e}{2}\,w_0' + H_3(\xi)w_1 + H_4(\xi)\frac{L_e}{2}\,w_1'$$

with the so-called Hermite-polynomials $H_i(\xi)$, $i = 1..4$. The latter are given by

$$H_1(\xi) = \tfrac{1}{4}(2 - 3\xi + \xi^3)\,,\ H_2(\xi) = \tfrac{1}{4}(1 - \xi - \xi^2 + \xi^3)$$
$$H_3(\xi) = \tfrac{1}{4}(2 + 3\xi - \xi^3)\,,\ H_4(\xi) = \tfrac{1}{4}(-1 - \xi + \xi^2 + \xi^3)$$

and depicted in Fig. 2.14.

The weak form does not only contain trial and test function but also their second derivatives. The result for the trial function reads

$$w_\square''(\xi) = \frac{1}{L_e^2}\left(6\xi\, w_0 + L_e(3\xi - 1)w_0' - 6\xi\, w_1 + L_e(3\xi + 1)w_1'\right)\,.$$

The second derivative of the test function is computed analogously. The results can be written compactly as

$$w_\square(\xi) = \underline{\boldsymbol{N}}^{\mathrm{T}}\,\underline{\hat{\boldsymbol{w}}}_\square \qquad v_\square(\xi) = \underline{\boldsymbol{N}}^{\mathrm{T}}\,\underline{\hat{\boldsymbol{v}}}_\square \tag{2.138}$$

$$w_\square''(\xi) = \underline{\boldsymbol{B}}^{\mathrm{T}}\,\underline{\hat{\boldsymbol{w}}}_\square \qquad v_\square''(\xi) = \underline{\boldsymbol{B}}^{\mathrm{T}}\,\underline{\hat{\boldsymbol{v}}}_\square \tag{2.139}$$

with

$$\underline{\hat{\boldsymbol{w}}}_\square^{\mathrm{T}} = \left[w_0\ w_0'\ w_1\ w_1'\right] \qquad \underline{\hat{\boldsymbol{v}}}_\square^{\mathrm{T}} = \left[v_0\ v_0'\ v_1\ v_1'\right]$$

and

$$\underline{N}^{\mathrm{T}} = \left[H_1(\xi)\ H_2(\xi)\tfrac{L_e}{2}\ H_3(\xi)\ H_4(\xi)\tfrac{L_e}{2}\right] \tag{2.140}$$

$$\underline{B}^{\mathrm{T}} = \frac{1}{L_e^2}\left[6\xi\ (3\xi-1)L_e\ -6\xi\ (3\xi+1)L_e\right]. \tag{2.141}$$

Performing the integration indicated in (2.137) using (2.138) and (2.139) yields

$$\begin{aligned}\mathcal{W}_{e_\square} &= \int\limits_{\xi=-1}^{+1} EI(\underline{B}^{\mathrm{T}}\,\hat{\underline{w}}_\square)\,(\underline{B}^{\mathrm{T}}\,\hat{\underline{v}}_\square))\,\frac{L_e}{2}\mathrm{d}\xi - \int\limits_{\xi=-1}^{+1} q\underline{N}^{\mathrm{T}}\,\hat{\underline{v}}_\square\,\frac{L_e}{2}\mathrm{d}\xi \\ &= \hat{\underline{v}}_\square^{\mathrm{T}}\left[\underline{\underline{\mathbf{K}}}_e\,\hat{\underline{w}}_\square - \underline{q}_e\right]\end{aligned} \tag{2.142}$$

with element stiffness matrix

$$\underline{\underline{K}}_e = \int\limits_{-1}^{+1} EI\,\underline{B}\,\underline{B}^{\mathrm{T}}\frac{L_e}{2}\mathrm{d}\xi \tag{2.143}$$

and element load vector

$$\underline{q}_e = \int\limits_{-1}^{+1} q\,\underline{N}\frac{L_e}{2}\mathrm{d}\xi\,. \tag{2.144}$$

Before inserting (2.142) in $\mathcal{W}_h$, the relation between the element "e" and the global node variables has to be encoded again. Analogously to Sect. 2.1.7 we use

$$\hat{\underline{w}}_\square \mapsto \underline{\underline{A}}_e\,\hat{\underline{w}} \qquad\qquad \hat{\underline{v}}_\square \mapsto \underline{\underline{A}}_e\,\hat{\underline{v}} \tag{2.145}$$

which eventually yields

$$\mathcal{W}_h = \hat{\underline{v}}^{\mathrm{T}}\left[\underline{\underline{K}}\,\hat{\underline{w}} - \underline{q} - \hat{\underline{F}}\right] = 0 \tag{2.146}$$

with

$$\underline{\underline{K}} = \left\{ \sum_{e=1}^{N} \underline{\underline{A}}_e^{\mathrm{T}} \underline{\underline{K}}_e \underline{\underline{A}}_e \right\} , \quad \underline{q}_e = \left\{ \sum_{e=1}^{N} \underline{\underline{A}}_e^{\mathrm{T}} \underline{q}_e \right\} \tag{2.147}$$

and

$$\begin{aligned} \underline{\hat{F}}^{\mathrm{T}} &= \left[V(0) \; -M(0) \; 0 \; 0 \ldots F \; -M_l \right] , \\ \underline{\hat{w}}^{\mathrm{T}} &= \left[w_1 \;\; w_1' \;\; w_2 \;\; w_2' \;\; \ldots \;\; w_{N_e+1} \;\; w_{N_e+1}' \right] , \\ \underline{\hat{v}}^{\mathrm{T}} &= \left[v_1 \;\; v_1' \;\; v_2 \;\; v_2' \;\; \ldots \;\; v_{N_e+1} \;\; v_{N_e+1}' \right] . \end{aligned}$$

Since arbitrariness of the test function means arbitrariness of $\underline{\hat{v}}$, Eq. (2.142) can only be fulfilled if

$$\underline{\underline{K}}\, \underline{\hat{w}} = \underline{q} + \underline{\hat{F}} . \tag{2.148}$$

The determinant of $\underline{\underline{K}}$ vanishes, which means, that the system of equations has infinitely many solutions and essential boundary conditions have to be incorporated in order to determine the solution for the considered problem.

In the following, the element stiffness matrix $\underline{\underline{K}}_e$ is computed for constant EI According to (2.147) and (2.141), we obtain

$$\underline{\underline{K}}_e = \int\limits_{\xi=-1}^{+1} \frac{1}{L_e^4} \begin{bmatrix} 6\xi \\ [3\xi - 1]L_e \\ -6\xi \\ [3\xi + 1]L_e \end{bmatrix} \left[6\xi \;\; [3\xi - 1]L_e \;\; -6\xi \;\; [3\xi + 1]L_e \right] EI \frac{L_e}{2} \mathrm{d}\xi$$

which is a quadratic matrix of dimension four. The integration has to be performed for every element of the matrix. Exemplary, we compute the element $K_{e(13)}$

$$K_{e(13)} = \frac{EI}{2L_e^3} \int\limits_{-1}^{+1} 6\xi(-6\xi)\mathrm{d}\xi = -\frac{18EI}{L_e^3} \frac{\xi^3}{3} \Bigg|_{-1}^{+1} = -\frac{36}{3} \frac{EI}{L_e^3} = -12 \frac{EI}{L_e^3} . \tag{2.149}$$

The final result for $\underline{\underline{K}}_e$ reads as follows

$$\underline{\underline{K}}_e = \frac{EI}{L_e^3}\begin{bmatrix} 12 & 6L_e & -12 & 6L_e \\ 6L_e & 4L_e^2 & -6L_e & 2L_e^2 \\ -12 & -6L_e & 12 & -6L_e \\ 6L_e & 2L_e^2 & -6L_e & 4L_e^2 \end{bmatrix}. \tag{2.150}$$

The computation of $\underline{q}_e$ for $q = \text{const.}$ is left as an exercise to the reader. The result is

$$\underline{q}_e^{\mathrm{T}} = q\frac{L_e}{2}\left[\, 1 \;\; \tfrac{L_e}{6} \;\; 1 \;\; -\tfrac{L_e}{6} \,\right]. \tag{2.151}$$

Example 2.12 The finite element scheme is demonstrated for the problem shown in the figure below.

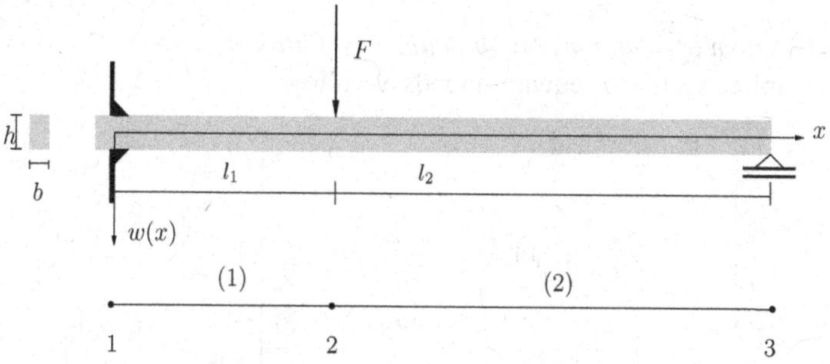

1. *Discretization*
   Two elements are used as shown in the figure together with the corresponding numbering of nodes and elements.
2. *Assembling element stiffness matrices and load vectors*
   Both element load vectors are zero because $q = 0$. The stiffness matrices are obtained as

$$\underline{\underline{K}}_1 = EI\begin{bmatrix} \frac{12}{l_1^3} & \frac{6}{l_1^2} & -\frac{12}{l_1^3} & \frac{6}{l_1^2} \\ & \frac{4}{l_1} & -\frac{6}{l_1^2} & \frac{2}{l_1} \\ & & \frac{12}{l_1^3} & -\frac{6}{l_1^2} \\ \text{symm} & & & \frac{4}{l_1} \end{bmatrix}, \quad \underline{\underline{K}}_2 = EI\begin{bmatrix} \frac{12}{l_2^3} & \frac{6}{l_2^2} & -\frac{12}{l_2^3} & \frac{6}{l_2^2} \\ & \frac{4}{l_2} & -\frac{6}{l_2^2} & \frac{2}{l_2} \\ & & \frac{12}{l_2^3} & -\frac{6}{l_2^2} \\ \text{symm} & & & \frac{4}{l_2} \end{bmatrix}$$

   with $I = \frac{b\,h^3}{12}$.
3. *Global stiffness matrix and global load vector*
   The gathering matrix for element one is derived from

$$e = 1: \begin{bmatrix} w_0 \\ w'_0 \\ w_1 \\ w'_1 \end{bmatrix} = \underbrace{\begin{bmatrix} 1\,0\,0\,0\,0\,0 \\ 0\,1\,0\,0\,0\,0 \\ 0\,0\,1\,0\,0\,0 \\ 0\,0\,0\,1\,0\,0 \end{bmatrix}}_{\underline{\underline{A}}_1} \begin{bmatrix} w_1 \\ w'_1 \\ w_2 \\ w'_2 \\ w_3 \\ w'_3 \end{bmatrix}$$

and determining the gathering matrix for element number two is left as an exercise to the reader. The global stiffness matrix is assembled according to

$$\underline{\underline{K}} = \underline{\underline{A}}_1^{\mathrm{T}} \underline{\underline{K}}_1 \underline{\underline{A}}_1 + \underline{\underline{A}}_2^{\mathrm{T}} \underline{\underline{K}}_2 \underline{\underline{A}}_2 \,.$$

4. *Global system of equations and boundary conditions*
The complete system of equation reads as follows

$$\underbrace{EI \begin{bmatrix} \frac{12}{l_1^3} & \frac{6}{l_1^2} & -\frac{12}{l_1^3} & \frac{6}{l_1^2} & 0 & 0 \\ & \frac{4}{l_1} & -\frac{6}{l_1^2} & \frac{2}{l_1} & 0 & 0 \\ & & \left(\frac{12}{l_1^3} + \frac{12}{l_2^3}\right) & \left(-\frac{6}{l_1^2} + \frac{6}{l_2^2}\right) & -\frac{12}{l_2^3} & \frac{6}{l_2^2} \\ & & & \left(\frac{4}{l_1} + \frac{4}{l_2}\right) & -\frac{6}{l^2} & \frac{2}{l_2^2} \\ & & & & \frac{12}{l_2^3} & -\frac{6}{l_2^2} \\ \text{symm} & & & & & \frac{4}{l_2^2} \end{bmatrix}}_{\underline{\underline{K}}} \underbrace{\begin{bmatrix} w_1 \\ w'_1 \\ w_2 \\ w'_2 \\ w_3 \\ w'_3 \end{bmatrix}}_{\hat{\underline{w}}} = \underbrace{\begin{bmatrix} V(0) \\ -M(0) \\ F \\ 0 \\ V(l) \\ 0 \end{bmatrix}}_{\hat{\underline{F}}} .$$

According to the problem and the discretization used, $w_1 = 0$, $w_2 = 0$ and $w'_1 = 0$. Therefore, the first two columns and the fifth column of $\underline{\underline{K}}$ can be deleted together with the corresponding entries in $\hat{\underline{w}}$. Furthermore, the corresponding rows can be deleted from the system because of the properties of the test function. The reduced system of equation eventually reads as follows

$$EI \begin{bmatrix} 12\left(\frac{1}{l_1^3} + \frac{1}{l_2^3}\right) & 6\left(-\frac{1}{l_1^2} + \frac{1}{l_2^2}\right) & \frac{6}{l_2^2} \\ & 4\left(\frac{1}{l_1} + \frac{1}{l_2}\right) & \frac{2}{l_2^2} \\ \text{symm} & & \frac{4}{l_2} \end{bmatrix} \begin{bmatrix} w_2 \\ w'_2 \\ w'_3 \end{bmatrix} = \begin{bmatrix} F \\ 0 \\ 0 \end{bmatrix} .$$

5. *Solution*

$$w_2 = \frac{F}{4EI} \frac{4l_1^3 l_2^3 + 3l_1^4 l_2^2}{3[l_1^3 + l_2^3] + 9[l_1 l_2^2 + l_1^2 l_2]}$$

$$w'_2 = \frac{F}{4EI} \frac{2l_1^2 l_2^3 - l_1^4 l_2}{4[l_1^3 + l_2^3] + 3[l_1 l_2^2 + l_1^2 l_2]}$$

$$w'_3 = -\frac{F}{4EI} \frac{l_1^2 l_2}{l_1 + l_2}$$

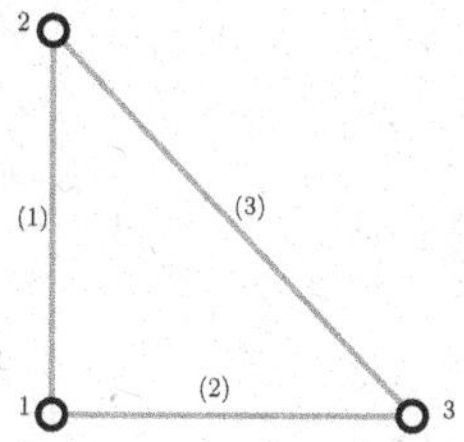

Fig. 2.15 Geometric realization of a simple graph with three nodes and three edges

6. *Accuracy estimation via natural boundary conditions* At the right hand side, the moment has to vanish. This can be used to check the accuracy of the solution. The moment at the right hand side corresponds to the moment in element number two at $\xi = +1$, i.e.,

$$M(l_1 + l_2) = -\,EI\,w_h''(x = l_1 + l_2) = -EI\,\underline{B}^{\mathrm{T}}(\xi = 1)\,\underline{\underline{A}}_2\,\underline{\hat{w}}$$

$$= -\,EI\begin{bmatrix} 6 & 2l_2 & -6 & 4l_2 \end{bmatrix}\begin{bmatrix} w_2 \\ w_2' \\ w_3 \\ w_3' \end{bmatrix}\frac{1}{l_2^2} = 0\,.$$

For the case of a concentrated load, the exact solution is found by FEM. The reason is the following. From the differential equation of the strong form of the boundary value problem, it can be seen, that for a concentrated force and constant beam stiffness, the exact solution for the moment depends only linearly on spatial location. The second derivative of the cubic polynomial used as approximation is a linear function. Therefore, the exact solution is obtained.

2.6 Network Models

The term network is used here to define the combination of a graph with a model from engineering or science. Typical examples are truss networks in mechanics, resistor networks in electrical engineering, and pipe networks in hydraulics. Graphs consist of nodes together with information about connections between nodes. A geometric representation of a simple graph is depicted in Fig. 2.15.

Here, only geometric realizations of graphs with straight edges are considered. In the following, networks are discussed which emerge from assigning models given in Table 2.1 to the edges of a graph's geometric realization.

All cases given in Table 2.1 are modelled by the same ordinary differential equation. But, the latter is the result of simplifying a partial differential equation which describes the corresponding spatially higher-dimensional problem. It depends on the

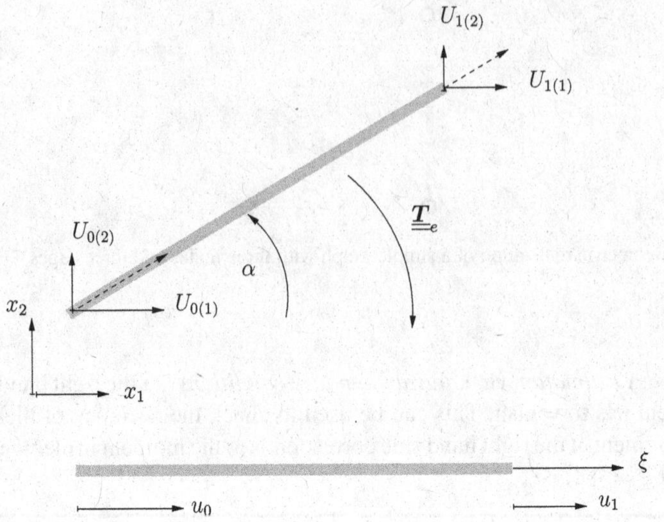

Fig. 2.16 Relation between the reference domain for axial deformation and an arbitrary finite element the plane

particular case, from which partial differential equation the spatially one-dimensional model is derived. For instance, linear transport is related to the Poisson equation (see Sect. 2.2) whereas elastic axial deformation is a special case of linear elasticity (see Sect. 2.4).

The type of the underlying partial differential equation determines the correct definition of corresponding network models.

The unknown displacement in the case of axial deformation is actually a vector. To assemble a planar network from individual parts, the geometric transformation of a finite element in the plane affects the vector of unknowns, as depicted in Fig. 2.16. The relation between the components of the displacement vectors at the nodes for an element inclined by an angle α and the displacements considering the reference domain Ω_e can be expressed in terms of a transformation matrix as follows

$$\underline{\hat{u}}_\square = \begin{bmatrix} u_0 \\ u_1 \end{bmatrix} = \begin{bmatrix} \cos\alpha & \sin\alpha & 0 & 0 \\ 0 & 0 & \cos\alpha & \sin\alpha \end{bmatrix} \begin{bmatrix} U_{0(1)} \\ U_{0(2)} \\ U_{1(1)} \\ U_{1(2)} \end{bmatrix} = \underline{\underline{T}}_e \, \underline{\hat{U}}_\square$$

which applies analogously for the values of the test function v, i.e.,

$$\underline{\hat{v}}_{\square} = \underline{\underline{T}}_e \, \underline{\hat{V}}_{\square} \, .$$

Therefore, (2.46) reads

$$\mathcal{W}_e = \underline{\hat{V}}^{\mathrm{T}} \left[\underline{\underline{A}}^{\mathrm{T}} \underline{\underline{K}}^*_e \underline{\underline{A}}_e \underline{\hat{U}} - \underline{\underline{A}}^{\mathrm{T}}_e \underline{q}^*_e \right] .$$

with

$$\underline{\underline{K}}^*_e = \underline{\underline{T}}^{\mathrm{T}}_e \underline{\underline{K}}_e \underline{\underline{T}}_e$$

and

$$\underline{q}^*_e = \underline{\underline{T}}^{\mathrm{T}}_e \underline{q}_e \, .$$

It should be noted, that for networks, the matrices $\underline{\underline{A}}_e$ have to be adopted accordingly. For constant stiffness λ^0 and constant line load n_0, element stiffness matrix $\underline{\underline{K}}^*_e$ and load vector $\underline{q}^*_e$ are given for an element with length l_e by

$$\underline{\underline{K}}^*_e = \frac{\lambda^0}{l_e} \begin{bmatrix} a & b & -a & -b \\ b & c & -b & -c \\ -a & -b & a & b \\ -b & -c & b & c \end{bmatrix} , \qquad \underline{q}^*_e = \frac{n_0 l_e}{2} \begin{bmatrix} \cos\alpha \\ \sin\alpha \\ \cos\alpha \\ \sin\alpha \end{bmatrix}$$

where $a = \cos^2\alpha$, $b = \sin\alpha\cos\alpha$ and $c = \sin^2$.

For linear transport like, for instance, Fourier heat conduction, the situation is completely different. Here, the unknown values at the nodes are temperatures. Since, temperature is a scalar quantity, it is not affected by a geometric transformation. Furthermore, $\underline{\hat{F}}$ contains thermal energies entering or leaving the system at the nodes. Hence $\underline{\hat{F}}$ is not affected neither by a geometric transformation.

2.7 Hand Calculation Exercises

2.8 Linear elastic frame element

A frame element is obtained by combining the Euler-Bernoulli beam with axial deformation. Derive the element stiffness matrix and the element load vector for a frame element in the plane inclined by an angle α, see, Fig. 2.16, for constant stiffness parameters in terms of axial deformation and bending.

2.9 Beam structure

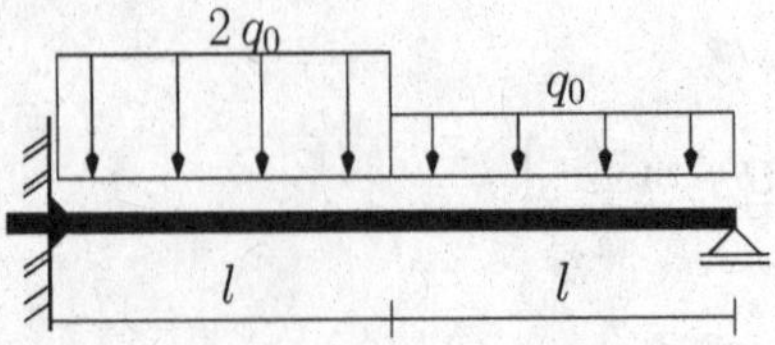

Given: q_0, l, EI constant

Determine approximations for $w(x)$, $w'(x)$ as functions of x and $M(x = l)$ with at most two finite elements.

2.8 Computer Exercises

Commercial software like Abaqus, Ansys, Sap2000, SiemensNX, or an Open Source platform such as Code Aster, Fenics, Feap, or similar, should be used to execute the exercises given in this section.

2.10 Truss structure

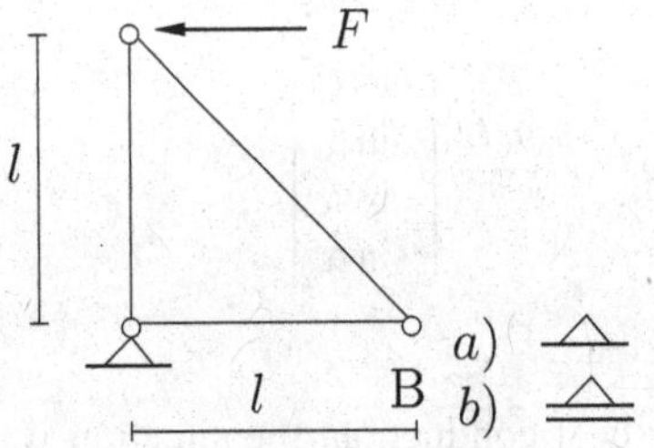

Given:

- F, $E = 200$ GPa, $A = 1256$ mm^2
- $l = 1$m

Determine the displacements at the nodes and all internal forces for

- support B with variant (a)
- support B with variant (b)

2.11 Linear elastic beam-like component

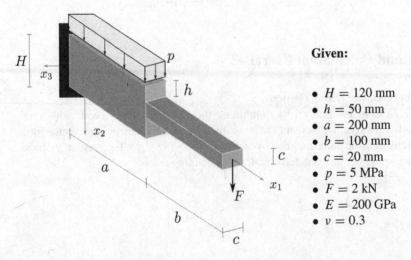

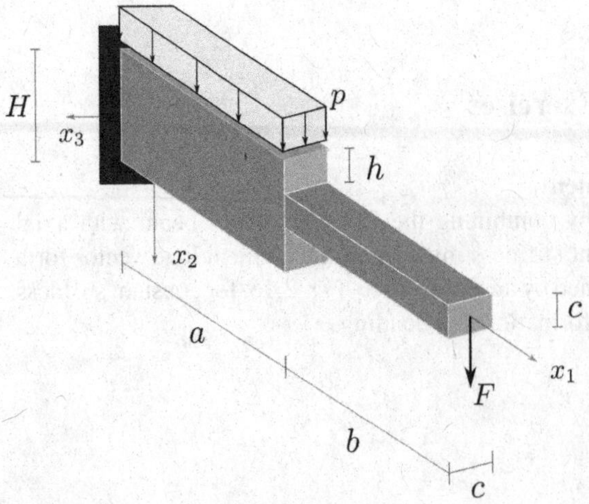

Given:

- $H = 120$ mm
- $h = 50$ mm
- $a = 200$ mm
- $b = 100$ mm
- $c = 20$ mm
- $p = 5$ MPa
- $F = 2$ kN
- $E = 200$ GPa
- $\nu = 0.3$

Execute the following tasks:

- Compute the vertical displacement along the line with coordinates $(x_1, 0, 0)$ employing a
  1. three-dimensional FE-model,
  2. two-dimensional FE-model with plane stress elements,
  3. one-dimensional FE-model with Euler-Bernoulli beam elements,
  4. one-dimensional FE-model with Timoshenko beam elements,

  and compare the results graphically.
- In addition, inspect normal and shear stresses for at least three different cross sections in view of the hypotheses used for deriving Euler-Bernoulli beam theory.
- For the two-dimensional model, inspect the dependence of the stress results on the finite element size in the region where the cross section height changes.

2.12 Linear elastic plate with circular hole

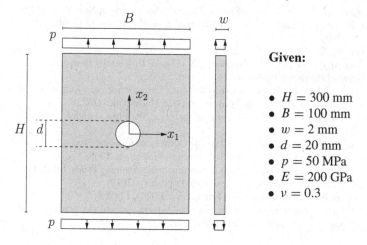

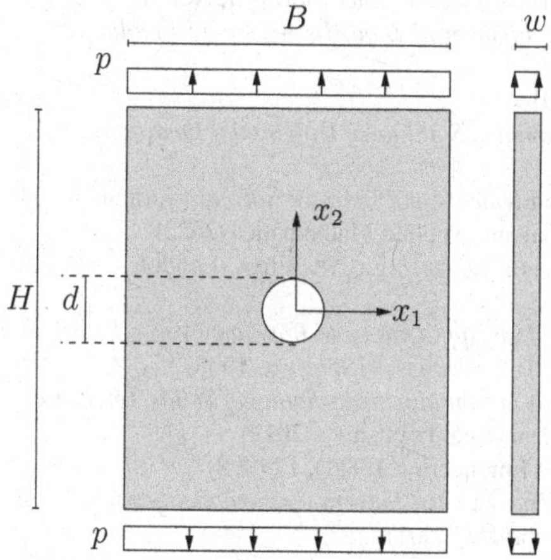

Given:

- $H = 300$ mm
- $B = 100$ mm
- $w = 2$ mm
- $d = 20$ mm
- $p = 50$ MPa
- $E = 200$ GPa
- $\nu = 0.3$

Design a mesh to assure sufficient accuracy of the stresses σ_{11} and σ_{22} along the path from $(x_1 = \frac{d}{2}, x_2 = 0)$ to $(x_1 = \frac{B}{2}, x_2 = 0)$.

2.9 Bibliographical Remarks

Sound knowledge about the underlying theory of a particular model is always helpful for discussing plausibility of numerical results. In terms of linear elastostatics, we recommend [2] and for the German speaking community [1].

Only the most common one and two dimensional finite elements are discussed in this chapter. Information about three dimensional elements and other finite element families can be found, for instance, in [3]. For more recent developments in Finite Element technology employing exterior algebra see, for instance, [4] or [5]. The latter offers in addition a comprehensive overview of the contemporary open source FEM platform Fenics. A detailed exposition of modern solver techniques can be found in [6].

Readers interested in the mathematical underpinnings of FEM in terms of Functional Analysis are referred to the introductory texts [7, 8]. A more advanced but recommendable text, which discusses methods of Functional Analysis for problems of Continuum Mechanics, is [9]. For additional sources, see the bibliographical remarks of Appendix C. Regarding error estimation methods based on Functional Analysis see, for instance, [10, 11], as well as references cited therein.

References

1. R. Kienzler, R. Schröder, *Einführung in die Höhere Festigkeitslehre* (Springer-Lehrbuch, Berlin, 2018)
2. P. Gould, *Introduction to Linear Elasticity* SpringerLink: Bücher (Springer, New York, 2013)
3. O. Zienkiewicz, R. Taylor, J. Zhu, *The Finite Element Method: Its Basis and Fundamentals* (Elsevier Science, 2005)
4. D.N. Arnold, A. Logg, SIAM News **47**(9) (2014)
5. A. Logg, K.A. Mardal, G.N. Wells, et al., *Automated Solution of Differential Equations by the Finite Element Method* (Springer, Berlin, 2012)
6. P. Gatto, *Mathematical Foundations of Finite Elements and Iterative Solvers*. Computational Science and Engineering (Society for Industrial and Applied Mathematics, 2022)
7. H. Shima, *Functional Analysis for Physics and Engineering* (CRC Press, London, England, 2016)
8. B. Reddy, *Introductory Functional Analysis: With Applications to Boundary Value Problems and Finite Elements*. Introductory Functional Analysis Series (Springer, 1998)
9. M. Kružík, T. Roubíček, *Mathematical Methods in Continuum Mechanics of Solids*. Interaction of Mechanics and Mathematics (Springer International Publishing, 2019)
10. Computer Methods in Applied Mechanics and Engineering **158**(1), 1 (1998)
11. D. Braess, L. Schumaker, *Finite Elements: Theory, Fast Solvers, and Applications in Solid Mechanics* (Cambridge University Press, Cambridge, 2001)

Chapter 3
Linear Initial Boundary Value Problems

Abstract After illustrating briefly the new aspects introduced by a time dependence, common time integration methods such as Euler's methods and Runge-Kutta methods are laid out. Stability of time integration schemes is discussed briefly. Main aspects of FEM solution schemes are illustrated for spatially one-dimensional problems, considering exemplary non-stationary linear transport and linear structural dynamics. In the context of the latter, Newmark's method is presented in more detail due to its popularity.

3.1 Introductory Example

To illustrate main aspects regarding the numerical solution of linear initial boundary value problems by means of FEM, the partial differential equation

$$\alpha \dot{u} + q' = r \quad x \in \Omega, t \in \tau \tag{3.1}$$

with

$$q = -\lambda u' \tag{3.2}$$

is considered, where $\Omega = (0, l)$ and $\tau = (0, t_{\mathrm{E}})$. Dot and prime indicate partial derivatives with respect to time t and spatial coordinate x, respectively.

Equations (3.1) and (3.2) can be condensed into

$$\alpha \dot{u} - (\lambda u')' = r \quad x \in \Omega, t \in \tau \tag{3.3}$$

Just like all other involved functions, u depends now in general on spatial location x and time t, i.e., $u = u(x, t)$. Therefore, information for $t = 0$ has to be provided in addition to the boundary conditions, which now can depend as well on time. Here the boundary conditions

$$u(0, t) = \overline{u}\, t \tag{3.4}$$
$$\lambda u'(l, t) = -\overline{q} \tag{3.5}$$

U. Mühlich, *Enhanced Introduction to Finite Elements for Engineers*, Solid Mechanics and Its Applications 268, https://doi.org/10.1007/978-3-031-30422-4_3

are assumed together with the initial condition

$$u(x, 0) = 0\,. \tag{3.6}$$

Remark 3.1 (*Interpretation of* (3.3)) Equation (3.3) can be used, for instance, to model heat transport in a continuous medium. In this case, q corresponds to the heat flux which depends linearly on the spatial gradient of the temperature u, (3.4) and (3.5) refer to a linear increase in temperature at $x = 0$ and a constant inward heat flux of magnitude $\overline{q}$ at $x = l$, respectively, whereas internal heat sources or sinks are taken into account by r. Furthermore, α denotes the product of specific heat capacity and density.

Approximating time derivatives properly requires a time discretisation. The time interval τ is split into disjoint time intervals $\tau_m = [t_m, t_{m+1})$ with lengths h_m given by $h_m = t_{m+1} - t_m$, such that

$$\tau = \bigcup_{m=0}^{M} \tau_m$$

where $t_{M+1} = \tau_{\mathrm{E}}$. The time derivative of u at a time t_{m+1} can be approximated, for instance, by

$$\dot{u}_{m+1} \approx \frac{1}{h_m}[u_{m+1} - u_m]\,. \tag{3.7}$$

For the sake of readability, explicit reference to the time t_{m+1} is omitted in the remaining part of this subsection with the understanding, that all time dependent quantities without explicit indication refer to $t = t_{m+1}$. With this notational convention, (3.7) can be written as

$$\dot{u} \approx \frac{1}{h_m}[u - u_m]\,. \tag{3.8}$$

In order for this scheme to work, all necessary information must be available at the beginning of a time interval also called time increment in the following. In consequence, (3.3) is evaluated at time t_{m+1} using (3.8), which yields the following strong form of the boundary value problem at time t_{m+1}

$$\begin{aligned} \alpha u - (\lambda u')' h_m - r h_m - \alpha u_m &= 0 \\ u(0) &= \overline{u}\, t_{m+1} \\ \lambda u'(l) &= -\,\overline{q} \end{aligned}$$

for given u_m and r. The functions u, v and r refer to time t_{m+1} and depend, therefore, only on x. This applies similarly to the boundary conditions.

The time-discrete variational form can now be derived in the same way as described in detail in Chap. 2. The result reads

$$\int_{\Omega} \lambda u' v' \,\mathrm{d}x + \int_{\Omega} \alpha u\, v \,\mathrm{d}x - \int_{\Omega} r\, v \,\mathrm{d}x - \int_{\Omega} \alpha u_m\, v \,\mathrm{d}x + \overline{q} h_m = 0$$

$$u(0) = \overline{u}\, t_{m+1}$$

which can now be solved using FEM as discussed in Chap. 2 and the solution determines the initial conditions for the next time interval.

Remark 3.2 The final result does not change if time discretisation (3.8) is applied to a time-continuous variational form obtained directly from (3.3), i.e., for this particular case, the order of steps can be reversed.

Several time integration methods exist besides (3.8). Time integration is actually a delicate issue because it may affect significantly accuracy and stability of the solution. This is discussed in more detail next.

3.2 Integration of Initial Value Problems

3.2.1 *General Aspects*

A function y which depends on time t is considered. The information about y is provided in terms of an ordinary differential equation

$$\dot{y} = f(t, y) \qquad t \in \tau = (t_0, t_\mathrm{E}) \tag{3.9}$$

with $f(t, y)$, for instance, $f(t, y) = \exp(\alpha t) y^2$, and the initial condition $y(t_0) = y_0$. Integrating both sides of the differential equation from t_0 to $t = t_{m+1}$ yields

$$y_{m+1} - y_0 = \int_{t_0}^{t_{m+1}} f \,\mathrm{d}t = \int_{t_0}^{t_m} f \,\mathrm{d}t + \int_{t_m}^{t_{m+1}} f \,\mathrm{d}t\,,$$

where, in view of the methods to be discussed, the domain of integration is split into two parts. Reordering gives

$$y_{m+1} = y_m + \int_{t_m}^{t_{m+1}} f \,\mathrm{d}t \tag{3.10}$$

by setting, in addition,

$$y_m = y_0 + \int_{t_0}^{t_m} f \, \mathrm{d}t \,.$$

If the integrals cannot be solved analytically, integration has to be performed numerically.

A large number of numerical integration methods exist. Information about the accuracy of a particular method is important for applying it properly. Estimates can be obtained by comparing corresponding Taylor series expansions with the Taylor series of the exact solution at $t = t_m$. For time steps of equal size h, the latter can be written as

$$y_{m+1} = y_m + f\,h + \frac{1}{2}\dot{f}\,h^2 + \frac{1}{6}\ddot{f}\,h^3 + \mathcal{O}(h^4) \tag{3.11}$$

if (3.9) is taken into account. The quantities f, $\dot{f}$ and $\ddot{f}$ in (3.11) are evaluated at $t = t_m$. In order to proceed, the time derivatives have to be specified. Since f depends on $y(t)$, f is interpreted as a function which depends on two variables, namely t and y, and chain rule is applied which yields

$$\dot{f} = \frac{\partial f}{\partial y}\frac{\mathrm{d}y}{\mathrm{d}t} + \frac{\partial f}{\partial t} = f_{,y} f + f_{,t} =: F \tag{3.12}$$

by using the usual short-hand notation for partial derivatives for the sake of readability. For the second time derivative

$$\ddot{f} = \dot{F} = (f_{,y})^{\bullet} f + f_{,y}\dot{f} + (f_{,t})^{\bullet} \tag{3.13}$$

$$= \left[f_{,yy}\frac{\mathrm{d}y}{\mathrm{d}t} + f_{yt} \right] f + f_{,y}F + \left[f_{,ty}\frac{\mathrm{d}y}{\mathrm{d}t} + f_{,tt} \right] = f_{,y}F + G \tag{3.14}$$

is obtained with

$$G = f_{,yy}f^2 + 2\,f_{,ty}f + f_{,tt} \,.$$

Example 3.1 (*Implicit differentiation*) The initial value problem $\dot{y} = \alpha \exp(\beta\,t)$ $y^2(t)$ with $y(0) = 1$ has the following solution

$$y(t) = \frac{\beta}{\beta + \alpha[1 - \exp(\beta t)]} \,.$$

The reader is invited to compare the results obtained by explicit and implicit differentiation.

Using (3.12) and (3.13), the Taylor series for the exact solution reads

$$y_{m+1} = y_m + f\,h + \frac{1}{2} F\,h^2 + \frac{1}{6}[f_{,y}F + G]\,h^3 + \mathcal{O}(h^4)\,. \tag{3.15}$$

Given some approximation $\tilde{y}_{m+1}$, (3.15) will be used in the following, together with the local truncation error

$$\Delta_h = y_{m+1} - \tilde{y}_{m+1} \tag{3.16}$$

to discuss exemplary the accuracy of some methods.

3.2.2 *Finite Difference Methods and Simple Euler Methods*

The principle idea of finite difference methods consists in approximating derivatives by difference quotients. Various possibilities exist and a systematic definition which provides simultaneously information regarding accuracy can be achieved using Taylor series expansions at $t = t_m$ for $t_m \rightarrow t_m + h$ and $t_m \rightarrow t_m - h$

$$y_{m+1} = y_m + \dot{y}_m h + \frac{1}{2}\ddot{y}_m h^2 + \frac{1}{6}\dddot{y}_m h^3 + \mathcal{O}(h^4)\,, \tag{3.17}$$

$$y_{m-1} = y_m - \dot{y}_m h + \frac{1}{2}\ddot{y}_m h^2 - \frac{1}{6}\dddot{y}_m h^3 + \mathcal{O}(h^4)\,. \tag{3.18}$$

The forward difference quotient can be obtained from (3.17) which requires to neglect terms of order h^2. Analogously, the backward difference quotient is obtained from (3.18). A better approximation can be achieved by means of the central difference quotient, obtained from the difference between (3.17) and (3.18) neglecting terms of order h^3. The results read as follows

$$\dot{y}_m \approx \frac{1}{h}\begin{cases} \left[y_{m+1} - y_m\right] & \text{forward} \\ \frac{1}{2}\left[y_{m+1} - y_{m-1}\right] & \text{central} \\ \left[y_m - y_{m-1}\right] & \text{backward} \end{cases} \quad \text{difference quotient}\,. \tag{3.19}$$

Adding (3.17) and (3.18) allows for defining an approximation for the second derivative at t_m

$$\ddot{y}_m \approx \frac{1}{h^2}\left[y_{m+1} - 2y_m + y_{m-1}\right]. \tag{3.20}$$

which requires to neglect terms of order h^4.

Exemplary, the finite difference method is applied to (3.9) by considering first (3.9) at time t_m and replacing $\dot{y}_m$ by the forward difference quotient. Eventually,

$$y_{m+1} = y_m + f(t_m, y_m)h$$

is obtained which allows for computing the solution at the end of every time step, based on its value at the beginning of the step. The procedure starts from an initial value defined via a suitable initial condition.

A common feature of time integration using finite difference methods is, that time derivatives are assumed to be constant within a time step. Simple Euler methods make this idea more explicit by replacing the function f in (3.10) by a constant, i.e., a polynomial of degree zero, for $t \in [t_m, t_{m+1}]$. Specifically,

$$f = c_0$$

is set in (3.10) and integration yields

$$\tilde{y}_{m+1} = y_m + c_0[t_{m+1} - t_m] = y_m + c_0 h\,. \tag{3.21}$$

The question arises, what to use for c_0. Possible choices are f_m and f_{m+1} but any value in between seems perfectly reasonable too. For $c_0 := f_m$

$$\tilde{y}_{m+1} = y_m + f_m h \tag{3.22}$$

is obtained from (3.21), whereas $c_0 := f_{m+1}$ yields

$$\tilde{y}_{m+1} = y_m + f_{m+1} h\,. \tag{3.23}$$

Although, (3.22) and (3.23) look similar, they are in fact completely different, which is illustrated for

$$f = [1 - \exp(-\alpha y)]$$

with some constant α. For this particular case, (3.22) reads

$$\tilde{y}_{m+1} = y_m + [1 - \exp(-\alpha y_m)]\,h\,, \tag{3.24}$$

whereas

$$\tilde{y}_{m+1} = y_m + [1 - \exp(-\alpha y_{m+1})]\,h\,. \tag{3.25}$$

is obtained from (3.23). While y_{m+1} is explicitly given by (3.24), (3.25) is a nonlinear equation for y_{m+1} which, in addition, has to be solved numerically. Therefore, (3.22) is called an explicit integration method, whereas (3.23) is an implicit integration method. Furthermore, (3.22) and (3.25) are known as well as Euler's explicit and Euler's implicit method, respectively.

In view of the example above, explicit integration methods seem to be the preferable choice but, in contrast with implicit methods, they are only conditionally stable which will be discussed in more detail in Sect. 3.3.

In addition, evaluating the difference between (3.22) and the Taylor series of the exact solution (3.15) yields a local truncation error (3.16) proportional to h^2 for Euler's explicit method. It means, that the accuracy of this method is rather low.

Any choice between the two limits f_m and f_{m+1} can be expressed concisely by the so-called θ-rule as

$$\tilde{y}_{m+1} = y_m + [1-\theta] f_m h + \theta f_{m+1} h \tag{3.26}$$

with $\theta \in [0, 1]$. Euler's explicit and Euler's implicit method are obtained from (3.26) by setting $\theta = 0$ and $\theta = 1$, respectively.

Remark 3.3 In some cases, it is not necessary to solve a nonlinear equation when applying Euler's implicit method. For instance, if $f = g(t)y$ with some function g which depends only on t,

$$\tilde{y}_{m+1} = \frac{y_m}{1 - g(t_{m+1})h}$$

is readily obtained from (3.23) because f depends only linearly on y.

3.2.3 Runge-Kutta Methods

Taking the principal idea of the methods discussed within the previous section further, Runge-Kutta methods replace the function f in (3.10) by an interpolation polynomial of degree k, $k \geq 1$. The corresponding integration method for given k is called Runge-Kutta method of order $k+1$.

Runge-Kutta methods of second order are obtained by replacing f in (3.10) by a polynomial of degree one, which interpolates linearly, for instance, between t_m and t_{m+1}, i.e.,

$$\tilde{f} = f_m \frac{t_{m+1} - t}{h} - f_{m+1} \frac{t_m - t}{h} .$$

Performing the integration gives eventually the implicit second order Runge-Kutta method

$$\tilde{y}_{m+1} = y_m + \frac{1}{2}[f_m + f_{m+1}]h , \tag{3.27}$$

which is knowns as well as trapezoidal rule. A corresponding second order explicit method can be obtained as follows. Computation of f_{m+1} requires some estimate for y_{m+1}, which can be obtained by using an explicit Runge-Kutta method of first order. Defining

$$k_1 = f(t_m, y_m) \tag{3.28}$$

$$k_2 = f(t_m + h, y_m + k_1 h), \tag{3.29}$$

an explicit counterpart of (3.27) reads

$$\tilde{y}_{m+1} = y_m + \left[\frac{1}{2}k_1 + \frac{1}{2}k_2\right] h. \tag{3.30}$$

The above scheme can be interpreted as a special case of a more general approach given by

$$k_1 = f(t_m, y_m) \tag{3.31}$$

$$k_2 = f(t_m + c_2 h, y_m + a_{21} k_1 h) \tag{3.32}$$

and

$$\tilde{y}_{m+1} = y_m + [b_1 k_1 + b_2 k_2] h. \tag{3.33}$$

Defining the method by means of (3.31)–(3.33) allows to address the question if the coefficients c_2, a_{21}, b_1 and b_2 can be chosen systematically on the basis of some criterion, for instance, in terms of accuracy.

Setting $\Delta t = c_2 h$ and $\Delta y = a_{21} k_1 h$, Taylor expansion of k_2 at $t = t_m$ and $y = y_m$ is straight forward

$$\begin{aligned} k_2(t_m + \Delta t, y_m + \Delta y) =& k_2(t_m, y_m) + \frac{\partial k_2}{\partial t}\Delta t + \frac{\partial k_2}{\partial y}\Delta y \\ &+ \frac{1}{2}\frac{\partial^2 k_2}{\partial t^2}\Delta t^2 + \frac{1}{2}\frac{\partial^2 k_2}{\partial y^2}\Delta y^2 + \mathcal{O}(h^3) \end{aligned}$$

where all derivatives of k_2 are evaluated at $t = t_m$ and $y = y_m$. Taking into account the meanings of k_1, k_2, Δt and Δy, the Taylor expansion can be written as

$$\begin{aligned} k_2(t_m + c_2 h, y_m + a_{21} k_1 h) =& f + f_{,t}\, c_2\, h + f_{,y}\, f\, a_{21}\, h \\ &+ \frac{1}{2} f_{,tt}\, c_2\, h^2 + \frac{1}{2} f_{,yy}\, f^2\, a_{21}^2\, h^2 + \mathcal{O}(h^3) \end{aligned}$$

and $\tilde{y}_{m+1}$ reads

$$\begin{aligned} \tilde{y}_{m+1} =\; & y_m + [b_1 + b_2] f\, h + \left[f_{,t} b_2 c_2 + f_{,y}\, f\, a_{21}\, b_2\right] h^2 \\ &+ \frac{1}{2}\left[f_{,tt} b_2 c_2^2 + f_{,yy}\, f^2\, a_{21}^2\, b_2\right] h^3 + \mathcal{O}(h^4). \end{aligned} \tag{3.34}$$

Comparison of (3.34) and (3.15) reveals, that in order to be consistent up to order h^2, which corresponds to a local truncation error proportional to h^3, the equations

$$b_1 + b_2 = 1 \qquad\qquad a_{21} b_2 = b_2 c_2 = \frac{1}{2}$$

have to be fulfilled. Because the system of equations is underdetermined, an infinite number of possible choices for the coefficients b_1, b_2, a_{21} and c_2 exist. Therefore, the choice made in (3.28)–(3.30) is reasonable but not unique.

Analogously, for a quadratic interpolation polynomial with interpolation points at t_m, $t_{m+\frac{1}{2}} = t_m + \frac{1}{2}h$ and t_{m+1},

$$y_{m+1} = y_m + \frac{1}{6}[f_m + 4 f_{m+\frac{1}{2}} + f_{m+1}]h\,, \tag{3.35}$$

is derived. However, (3.35) does not even provide an implicit method yet because there is just one equation for two unknowns.

There are various ways to proceed from (3.35). But, instead of exploring some or all of them, explicit integration is considered first and the scheme given by (3.31)–(3.33) is extended accordingly, which gives

$$\begin{aligned} k_1 &= f(t_m, y_m) \\ k_2 &= f(t_m + c_2 h, y_m + a_{21} k_1 h) \\ k_3 &= f(t_m + c_3 h, y_m + a_{31} k_1 h + a_{32} k_2 h) \end{aligned}$$

and

$$\tilde{y}_{m+1} = y_m + [b_1 k_1 + b_2 k_2 + b_3 k_3]h\,.$$

Reasonable choices for the involved coefficients can be justified again on the basis of accuracy. More specifically, consistence of the method up to the order h^3 can be achieved by choosing the coefficients properly based on comparing y_{m+1} and $\tilde{y}_{m+1}$. The latter requires Taylor series expansions of k_2 and k_3, which is omitted here. The result is again an underdetermined system of nonlinear equations and, therefore, different reasonable choices for the coefficients are possible.

The result can be given concisely by corresponding matrices $\underline{\underline{A}}$, $\underline{b}$ and $\underline{c}$. The so-called Butcher tableau is commonly used to summarise the necessary information related to a specific method. With the help of the latter, a popular version of the explicit second order Runge-Kutta method can be summarised as

$$\begin{array}{c|c} \underline{c} & \underline{\underline{A}} \\ \hline & \underline{b}^{\mathrm{T}} \end{array} \quad\Rightarrow\quad \begin{array}{c|ccc} c_1 & a_{11} & & \\ c_2 & a_{21} & a_{22} & \\ c_3 & a_{31} & a_{32} & a_{33} \\ \hline & b_1 & b_2 & b_3 \end{array} \quad\Rightarrow\quad \begin{array}{c|ccc} 0 & 0 & & \\ \frac{1}{2} & \frac{1}{2} & 0 & \\ 1 & -1 & 0 & 1 \\ \hline & \frac{1}{6} & \frac{2}{3} & \frac{1}{6} \end{array}$$

indicating simultaneously the logic behind the Butcher tableau.

Runge-Kutta methods of order s can be cast into the following unified representation

$$k_i = f\left(t_m + c_i h,\ y_m + h\sum_{j=1}^{s} a_{ij}k_j\right) \tag{3.36}$$

$$\tilde{y}_{m+1} = y_m + h\sum_{i=1}^{s} b_i k_i\,.$$

For explicit methods

$$a_{ij} = 0 \quad \text{if } j > i$$

holds and, therefore, the k_i can be computed explicitly starting from k_1. For implicit methods, the matrix $\underline{\underline{A}}$ is not triangular but in general fully occupied. The most popular Runge-Kutta method is the fourth order explicit method, given usually by the following Butcher tableau

$$\begin{array}{c|cccc} 0 & 0 & & & \\ \frac{1}{2} & \frac{1}{2} & 0 & & \\ \frac{1}{2} & 0 & \frac{1}{2} & 0 & \\ 1 & 0 & 0 & 1 & 0 \\ \hline & \frac{1}{6} & \frac{1}{3} & \frac{1}{3} & \frac{1}{6} \end{array}\,.$$

Finally, it should be noted, that the coefficients b_i should always sum up to one, to ensure, that constant functions are integrated exactly.

3.3 Stability of Time Integration Schemes

General aspects regarding the stability of integration methods are illustrated by means of the example

$$\dot{y}(t) + Ky(t) = g(t)\,, \tag{3.37}$$

$$y(t=0) = y_0\,. \tag{3.38}$$

The solution depends on the initial condition y_0. In the following, the two cases

$$\begin{aligned} \dot{y}^{\mathrm{A}}(t) + Ky^{\mathrm{A}}(t) &= g(t) \\ y^{\mathrm{A}}(0) &= y_0^{\mathrm{A}} \end{aligned} \qquad\qquad \begin{aligned} \dot{y}^{\mathrm{B}}(t) + Ky^{\mathrm{B}}(t) &= g(t) \\ y^{\mathrm{B}}(0) &= y_0^{\mathrm{B}} \end{aligned}$$

are considered, where

$$r_0 = y^{\mathrm{A}}(0) - y^{\mathrm{B}}(0)$$

simulates an error in the initial conditions at time $t = 0$. The initial value problem that results from the difference of the two systems A and B

$$\dot{r}(\tau) = f(t, r) \tag{3.39}$$

$$r(0) = r_0 \tag{3.40}$$

with $r(t) = y^{\mathrm{A}}(t) - y^{\mathrm{B}}(t)$ and $f(t, r) = -Kr(t)$, models the evolution of the error $r(t)$ due to some error in the initial condition.

The evolution model (3.39) can be used to discuss the stability of numerical integration methods, which is demonstrated in the following for the θ-formalism (3.26). Applying the θ-formalism to (3.39) gives

$$r_{m+1} = r_m - [1-\theta]Kr_m\, h - \theta K\, r_{m+1}h\,, \tag{3.41}$$

from which

$$r_{m+1} = \frac{1-[1-\theta]K\,h}{1+K\theta\,h}\, r_m \tag{3.42}$$

can be obtained. Defining

$$Z = \frac{1-[1-\theta]K\,h}{1+K\theta\,h}$$

and evaluating r at different times yields the following recursive scheme

$$\begin{aligned} r_1 &= Zr_0 \\ r_2 &= Zr_1 = Z^2 r_0 \\ r_3 &= Zr_2 = Z^3 r_0 \\ &\vdots \end{aligned}$$

which can be cast into

$$r_m = Z^m r_0\,.$$

Depending on whether $|Z| > 1$, $|Z| = 1$ or $|Z| < 1$, the error increases, remains unchanged or decreases, respectively, i.e, the system is unstable, indifferent, or stable. It turns out, that the θ-formalism is stable if $\theta \geq \frac{1}{2}$, which implies, that Euler's implicit method is unconditionally stable.

An equally important questions is the following. Under which conditions, the θ-formalism is stable if $\theta < \frac{1}{2}$ and the result is $h \leq 2/K$. The corresponing calculations, which require case differentiation regarding the absolute value of Z, are left as an exercise to the reader.

3.4 Non-stationary Linear Transport

3.4.1 Time-Continuous Variational Form

A typical model for non-stationary transport in one spatial direction is (3.3), see, as well Remark 3.1. For the benefit of the reader, (3.3) is reiterated together with the complete set of initial and boundary conditions

$$\alpha \dot{u} - (\lambda u')' = r \qquad x \in \Omega, t \in \tau = [0, t_{\mathrm{E}}] \qquad \text{(3.3 revisited)}$$
$$u(0, t) = \overline{u}\, t \qquad \text{(3.4 revisited)}$$
$$\lambda u'(l, t) = -\overline{q} \qquad \text{(3.5 revisited)}$$
$$u(x, 0) = 0. \qquad \text{(3.6 revisited)}$$

Again, please note, that for the system given above, prime and dot are partial derivatives because all involved functions depend in general on spatial location and time.

Multiplying (3.3) with a test function v, which vanishes at $x = 0$ due to the presence of the essential boundary condition, and applying integration by parts yields after some manipulation the time continuous variational form

$$\mathcal{W} = \int_{\Omega} \alpha \dot{u}\, v \,\mathrm{d}x + \int_{\Omega} \lambda u'\, v' \,\mathrm{d}x - \int_{\Omega} r\, v \,\mathrm{d}x + \overline{q} v(l) = 0 \tag{3.43}$$

$$u(0, t) = \overline{u}\, t$$
$$u(x, 0) = 0$$

with the natural boundary condition incorporated as usual into (3.43).

There are various ways to proceed. Time discretisation of (3.43) and applying some time integration method yields a time discrete weak form for every time interval for which the methods of Functional Analysis provided in Appendix C apply. Equivalently, the time discrete variational form can be obtained by starting from a time discrete strong form, see Sect. 3.1.

Alternatively, spatial discretisation in terms of FEM can be performed for (3.43), assigning eventually the time dependence to the values of the unknown functions at the nodes. The result is a system of equation for the node values and their time derivatives for which a time integration method has to be applied.

3.4.2 Galerkin FEM

Within this section, the Galerkin method described in detail in Sect. 2.1.7 is applied for (3.43). Assuming a reasonable spatial discretisation, integrals in (3.43) are evaluated piecewise, i.e.,

$$\mathcal{W} = \sum_{e=1}^{N_e} \mathcal{W}_e + \overline{q}v(l) = 0 \tag{3.44}$$

with

$$\mathcal{W}_e = \int_{\Omega_e} \alpha \dot{u}\, v\, \mathrm{d}x + \int_{\Omega_e} \lambda u'\, v'\, \mathrm{d}x - \int_{\Omega_e} r\, v\, \mathrm{d}x\,.$$

A reference domain $\Omega_\square$ and corresponding mappings χ_e are defined. Details of the method are not reiterated because it is assumed, that the reader has already worked through Sect. 2.1.7.

To account for the time dependence of the problem, the definitions (2.37) and (2.38) are adjusted as follows

$$\begin{aligned} u_\square(\xi, t) &= \underline{\boldsymbol{N}}^{\mathrm{T}}\, \underline{\hat{\boldsymbol{u}}}_\square \\ u'_\square(\xi, t) &= \underline{\boldsymbol{B}}^{\mathrm{T}}\, \underline{\hat{\boldsymbol{u}}}_\square\, j_e^{-1} \end{aligned}$$

assuming, that the time dependence is taken into account by the nodal values. More specifically, contrary to Sect. 2.1.7, nodal values are no longer just real numbers but functions of time. Therefore, a time derivative affects only the nodal values, i.e.,

$$\dot{u}_\square(\xi, t) = \underline{\boldsymbol{N}}^{\mathrm{T}}\, \dot{\underline{\hat{\boldsymbol{u}}}}_\square\,,$$

whereas the spatial derivative still affects only the form functions. Recall, that j_e is the Jacobian of the mapping χ_e, $\underline{\boldsymbol{N}}$ contains the form functions, and $\underline{\boldsymbol{B}}$ their derivatives with respect to ξ.

The definition of the test functions with reference to $\Omega_\square$ remains unchanged and

$$\begin{aligned} v_\square(\xi) &= \underline{\boldsymbol{N}}^{\mathrm{T}}\, \underline{\hat{\boldsymbol{v}}}_\square && \text{(2.39 revisited)} \\ v'_\square(\xi) &= \underline{\boldsymbol{B}}^{\mathrm{T}}\, \underline{\hat{\boldsymbol{v}}}_\square\, j_e^{-1} && \text{(2.40 revisited)} \end{aligned}$$

are reiterated here for the benefit of the reader.

The contribution of a domain Ω_e to (3.44), expressed in terms of quantities defined for $\Omega_\square$, reads

$$\mathcal{W}_{e\square} = \underline{\hat{\boldsymbol{v}}}_\square^{\mathrm{T}} \left[\underline{\underline{\boldsymbol{S}}}_e\, \dot{\underline{\hat{\boldsymbol{u}}}}_\square + \underline{\underline{\boldsymbol{K}}}_e\, \underline{\hat{\boldsymbol{u}}}_\square - \underline{\boldsymbol{q}}_e \right] \tag{3.45}$$

with

$$\underline{\underline{S}}_e = \int_{-1}^{+1} \alpha \, \underline{N} \, \underline{N}^{\mathrm{T}} \frac{1}{j_e} \, \mathrm{d}\xi \tag{3.46}$$

and stiffness matrix $\underline{\underline{K}}_e$, as well as volume load vector $\underline{q}_e$, as given in Sect. 2.1.7. For constant α, $\underline{\underline{S}}_e$ reads

$$\underline{\underline{S}}_e = \frac{\alpha}{3 j_e} \begin{bmatrix} 2 & 1 \\ 1 & 2 \end{bmatrix}.$$

Taking into account the relations between local and global nodal values, for instance by using gathering matrices, leads eventually the following discrete version of (3.44)

$$\mathcal{W}_h = \underline{\hat{v}}^{\mathrm{T}} \left[\underline{\underline{S}} \, \dot{\underline{\hat{u}}} + \underline{\underline{K}} \, \underline{\hat{u}} - \underline{q} + \underline{\hat{F}} \right] = 0$$

with

$$\underline{\hat{F}}^{\mathrm{T}} = [q(0) \; 0 \; 0 \ldots 0 \; 0 \; \overline{q}\,.]$$

From the arbitrariness of the test function follows, that the following system of equations

$$\underline{\underline{S}} \, \dot{\underline{\hat{u}}} + \underline{\underline{K}} \, \underline{\hat{u}} = \underline{q} - \underline{\hat{F}}$$

has to be fulfilled. Before proceeding further, essential boundary conditions have to be incorporated, by deleting equations and matrix columns accordingly, which yields the constrained system of equations. Non-homogeneous Dirichlet boundary conditions require in addition some reorganisation. After this procedure, the system takes the form

$$\underline{\underline{S}} \, \dot{\underline{\hat{u}}} + \underline{\underline{K}} \, \underline{\hat{u}} = \underline{q} - \underline{\hat{F}} - \underline{\Gamma}^{\mathrm{D}} \tag{3.47}$$

where $\underline{\Gamma}^{\mathrm{D}}$ accounts for non-homogeneous Dirichlet boundary conditions, see, Example 3.2. For the sake of readability, unconstrained and constrained system are not distinguished further by notation. Instead, the specific meaning is indicated by the corresponding text if necessary.

Example 3.2 (*Problem* (3.44) *with two linear Lagrange elements*) The initial value problem given by (3.44) is considered with $\lambda = \lambda_0$ and $r = 0$. A spatial discretisation with two finite elements of equal length is used. Nodes and elements are numbered as in Example 2.9. The initial system of equations reads

$$\frac{\alpha}{3j_e}\begin{bmatrix} 2 & 1 & 0 \\ 1 & 4 & 1 \\ 0 & 1 & 2 \end{bmatrix}\begin{bmatrix} \dot{u}_1 \\ \dot{u}_2 \\ \dot{u}_3 \end{bmatrix} + \frac{2\lambda_0}{j_e}\begin{bmatrix} 1 & -1 & 0 \\ -1 & 2 & -1 \\ 0 & -1 & 1 \end{bmatrix}\begin{bmatrix} u_1 \\ u_2 \\ u_3 \end{bmatrix} = \begin{bmatrix} q(0) \\ 0 \\ \overline{q} \end{bmatrix}.$$

The essential boundary condition (3.4) corresponds to $u_1 = \overline{u}\,t$ and implies $\dot{u}_1 = \overline{u}$. Reordering the system accordingly gives

$$\frac{\alpha}{3j_e}\begin{bmatrix} 2 & 1 & 0 \\ 1 & 4 & 1 \\ 0 & 1 & 2 \end{bmatrix}\begin{bmatrix} 0 \\ \dot{u}_2 \\ \dot{u}_3 \end{bmatrix} + \frac{2\lambda_0}{j_e}\begin{bmatrix} 1 & -1 & 0 \\ -1 & 2 & -1 \\ 0 & -1 & 1 \end{bmatrix}\begin{bmatrix} 0 \\ u_2 \\ u_3 \end{bmatrix} = -\overline{u}\frac{\alpha}{3j_e}\begin{bmatrix} 2 \\ 1 \\ 0 \end{bmatrix} - \overline{u}t\frac{\lambda}{2j_e}\begin{bmatrix} 1 \\ -1 \\ 0 \end{bmatrix} + \begin{bmatrix} q(0) \\ 0 \\ \overline{q} \end{bmatrix}.$$

Since $v_1 = 0$, the first equation can be deleted. In addition, all matrix rows multiplied by zero can be deleted as well. Eventually

$$\begin{bmatrix} \dot{u}_2 \\ \dot{u}_3 \end{bmatrix} = -\frac{3j_e}{\alpha}\begin{bmatrix} 2 & -1 \\ -1 & 2 \end{bmatrix}\left\{\frac{2\lambda_0}{j_e}\begin{bmatrix} 2 & -1 \\ -1 & 1 \end{bmatrix}\begin{bmatrix} u_2 \\ u_3 \end{bmatrix} + \overline{u}\frac{\alpha}{3j_e}\begin{bmatrix} 1 \\ 0 \end{bmatrix} + \overline{u}\,t\frac{\lambda_0}{2j_e}\begin{bmatrix} -1 \\ 0 \end{bmatrix} - \begin{bmatrix} 0 \\ \overline{q} \end{bmatrix}\right\}$$

is obtained by isolating $\dot{\hat{\underline{u}}}$ via inversion.

3.4.3 *Time Integration*

Solving the constrained system (3.47) requires a time integration method. Evaluating (3.47) at time t_m and approximating all time derivatives in $\dot{\underline{u}}$ by forward difference quotients, see (3.19), is equivalent to Euler's forward method

$$\hat{\underline{u}}_{m+1} = \underline{\underline{S}}^{-1}\left[-\underline{\underline{K}}\,\hat{\underline{u}}_m + \underline{q}_m - \hat{\underline{F}}_m - \underline{\Gamma}^{\mathrm{D}}_m\right]h + \hat{\underline{u}}_m. \tag{3.48}$$

Similarly, evaluating (3.47) at time t_{m+1} and approximating all time derivatives in $\dot{\underline{u}}$ by backward difference quotients, see (3.19), yields the Euler backward scheme for the considered case

$$\hat{\underline{u}}_{m+1} = \left[\underline{\underline{S}} + h\underline{\underline{K}}\right]^{-1}\left[\underline{q}_{m+1} - \hat{\underline{F}}_{m+1} - \underline{\Gamma}^{\mathrm{D}}_{m+1}\right]h + \underline{\underline{S}}\,\hat{\underline{u}}_m. \tag{3.49}$$

Since $\underline{\underline{S}}$ and $\underline{\underline{K}}$ do not depend on time, they do not change during the time integration procedure. Therefore, their computation and inversion is required only once at the beginning of the procedure.

Applying Runge-Kutta methods, requires first to generalise (3.36) for systems of initial value problems, which is fortunately straight forward and the result reads

$$\underline{k}_i = \underline{f}\left(t_m + c_i h, \underline{\hat{u}}_m + h\sum_{j=1}^{s} a_{ij}\underline{k}_j\right)$$

$$\underline{\hat{u}}_{m+1} = \underline{\hat{u}}_m + h\sum_{i=1}^{s} b_i \underline{k}_i$$

if the vectors containing the unknown functions are replaced by the vectors of nodal displacements.

The vector of nodal velocities $\dot{\underline{\hat{u}}}$ in the constrained system (3.47) can be isolated by inversion, which yields

$$\dot{\underline{\hat{u}}} = -\underline{\underline{S}}^{-1}\left[\underline{\underline{K}}\,\underline{\hat{u}} + \underline{q} - \underline{\hat{F}} - \underline{\Gamma}^{\mathrm{D}}\right], \tag{3.50}$$

or, written more concisely,

$$\dot{\underline{\hat{u}}} = \underline{f}(t, \underline{\hat{u}}). \tag{3.51}$$

In the following, a second order explicit Runge-Kutta method is applied for the introductory example provided in Sect. 3.1 assuming $r = 0$. Due to the inhomogeneous Dirichlet boundary condition (3.4), $\underline{\Gamma}^{\mathrm{D}}$ reads

$$\underline{\Gamma}^{\mathrm{D}} = \underline{S}_1\overline{u} + \underline{K}_1\overline{u}\,t\,,$$

where $\underline{S}_1$ and $\underline{K}_1$ are the first columns of the matrices $\underline{\underline{S}}$ and $\underline{\underline{K}}$ of the unconstrained system after deleting the first entries, see, as well, Example 3.2. Furthermore, defining

$$\begin{aligned}
\underline{\underline{A}}^{\mathrm{SK}} &= \underline{\underline{S}}^{-1}\,\underline{\underline{K}}\,,\\
\underline{s}_1 &= \underline{\underline{S}}^{-1}\underline{\hat{F}}\,,\\
\underline{s}_2 &= \underline{\underline{S}}^{-1}\underline{S}_1\overline{u}\,,\\
\underline{s}_3 &= \underline{\underline{S}}^{-1}\underline{K}_1\overline{u}\,,
\end{aligned}$$

$\underline{f}(t, \underline{\hat{u}})$ can be written as

$$\underline{f}(t, \underline{\hat{u}}) = -\underline{\underline{A}}^{\mathrm{SK}}\underline{\hat{u}} - \underline{s}_1 + \underline{s}_2 + \underline{s}_3\,t\,. \tag{3.52}$$

Exemplary, a simple algorithm for implementing the second order explicit Runge-Kutta time integration for (3.52) is given as pseudocode in Algorithm 1.

Algorithm 1 Explicit second order Runge-Kutta for non-stationary transport

Require: $N_{\text{intervals}}, t_E, b_1, b_2, c_2, a_{21} > 0; \underline{\underline{A}}^{SK}, \underline{s}_1, \underline{s}_2, \underline{s}_3$
Ensure: $\hat{\underline{u}}$ for $t = t_E$

$h \leftarrow t_E/N_{\text{intervals}}$
$t \leftarrow 0$
$\hat{\underline{u}}_{m+1} \leftarrow \underline{0}$

while $t < t_E$ **do**

$\hat{\underline{u}}_m \leftarrow \hat{\underline{u}}_{m+1}$

$\underline{k}_1 \leftarrow -\underline{\underline{A}}^{SK}\hat{\underline{u}}_m + \underline{s}_1 + \underline{s}_2 + \underline{s}_3 t$
$\underline{k}_2 \leftarrow -\underline{\underline{A}}^{SK}\left[\hat{\underline{u}}_m + a_{21}\underline{k}_1 h\right] + \underline{s}_1 + \underline{s}_2 + \underline{s}_3[t + c_3 h]$

$\hat{\underline{u}}_{m+1} \leftarrow \hat{\underline{u}}_m + [b_1\underline{k}_1 + b_2\underline{k}_2]h$

$t \leftarrow t + h$

end while

3.5 Linear Structural Dynamics

3.5.1 Strong Form and Time-Continuous Variational Form

Considering only one spatial dimension, the strong form for structural dynamics reads

$$\rho\ddot{u} + \alpha\dot{u} - (\lambda u')' - r = 0 \quad x \in \Omega = (0, l) \tag{3.53}$$

with density ρ, damping function α and stiffness λ. It is a linear partial differential equation as long as ρ, α and λ do not depend on the displacement u or its time derivatives. Since (3.53) is second order in time and also second order in x, two initial conditions and two boundary conditions are required. Exemplary, here the boundary conditions

$$u(t, 0) = 0 \tag{3.54}$$

$$\lambda u'(t, l) = P_l(t) \tag{3.55}$$

with given function $P_l(t)$ and the initial conditions

$$u(0, x) = 0 \tag{3.56}$$

$$\dot{u}(0, x) = 0 \tag{3.57}$$

are used for illustration purposes.

The time-continuous variational form is derived analogously to the previous sections. The details are omitted here. The result reads

$$\int_{\Omega} \rho \ddot{u}\, v\, \mathrm{d}x + \int_{\Omega} \alpha \dot{u}\, v\, \mathrm{d}x + \int_{\Omega} \lambda u' v' \mathrm{d}x - \int_{\Omega} r\, v\, \mathrm{d}x - P_l(t) v(l) = 0 \qquad (3.58)$$

$$u(t, 0) = 0$$
$$\dot{u}(0, x) = 0$$
$$\ddot{u}(0, x) = 0$$

and the corresponding weak form is obtained by interpreting all spatial derivatives as weak derivatives.

3.5.2 Galerkin FEM and Time Integration with Central Differences

Linear Lagrange elements are used exemplary. Therefore, FEM discretisation follows the same steps as in Sect. 3.4.2.

The contribution of a domain Ω_e to (3.58), expressed in terms of quantities defined for $\Omega_\square$, reads

$$\mathcal{W}_{e_\square} = \underline{\hat{v}}_\square^{\mathrm{T}} \left[\underline{\underline{M}}_e\, \underline{\ddot{\hat{u}}}_\square + \underline{\underline{S}}_e\, \underline{\dot{\hat{u}}}_\square + \underline{\underline{K}}_e\, \underline{\hat{u}}_\square - \underline{q}_e \right] \qquad (3.59)$$

with element stiffness matrix $\underline{\underline{K}}_e$ and volume load vector $\underline{q}_e$, as given in Sect. 2.1.7. Although, $\underline{\underline{S}}_e$ refers now to damping, it has the exact same structure as given in Sect. 3.4.2. The only additional term is due to the inertia term in (3.58) and the element mass matrix is given by

$$\underline{\underline{M}}_e = \int_{-1}^{+1} \rho\, \underline{N}\, \underline{N}^{\mathrm{T}} \frac{1}{j_e}\, \mathrm{d}\xi \qquad (3.60)$$

For constant ρ, $\underline{\underline{M}}_e$ reads

$$\underline{\underline{M}}_e = \frac{\rho}{3 j_e} \begin{bmatrix} 2 & 1 \\ 1 & 2 \end{bmatrix}.$$

Taking into account the relations between local and global nodal values, for instance by using gathering matrices, leads eventually the following unconstrained system of equations

$$\underline{\underline{M}}\, \underline{\ddot{\hat{u}}} + \underline{\underline{S}}\, \underline{\dot{\hat{u}}} + \underline{\underline{K}}\, \underline{\hat{u}} = \underline{q} + \underline{\hat{F}} \qquad (3.61)$$

from which the constrained system

$$\underline{\underline{M}}\,\ddot{\hat{\underline{u}}} + \underline{\underline{S}}\,\dot{\hat{\underline{u}}} + \underline{\underline{K}}\,\hat{\underline{u}} = \underline{q} + \hat{\underline{F}} - \underline{\Gamma}^{\mathrm{D}} \tag{3.62}$$

is obtained by incorporating essential boundary conditions. Non-homogeneous Dirichlet boundary conditions are accounted for by $\underline{\Gamma}^{\mathrm{D}}$. For the details see, for instance, Example 3.2. Evaluating the constrained system at time $t = t_m$, approximating first time derivatives by means of central differences given in (3.19) and second derivatives using the difference quotient given by (3.20), gives

$$\underline{\underline{M}}\frac{1}{h^2}\left[\hat{\underline{u}}_{m+1} - 2\hat{\underline{u}}_m + \hat{\underline{u}}_{m-1}\right] + \underline{\underline{S}}\frac{1}{2h}\left[\hat{\underline{u}}_{m+1} + \hat{\underline{u}}_{m-1}\right] + \underline{\underline{K}}\,\hat{\underline{u}}_m = \underline{q}_m + \hat{\underline{F}}_m - \underline{\Gamma}^{\mathrm{D}}_m$$

and, after rearranging

$$\underline{\underline{\kappa}}\,\hat{\underline{u}}_{m+1} = 2\underline{\underline{M}}\,\hat{\underline{u}}_m - \left[\underline{\underline{M}} - \frac{h}{2}\underline{\underline{S}}\right]\hat{\underline{u}}_{m-1} + \underline{q}_m h^2 + \hat{\underline{F}}_m h^2 - \underline{\Gamma}^{\mathrm{D}} h^2 \tag{3.63}$$

with

$$\underline{\underline{\kappa}} = \underline{\underline{M}} + \frac{h}{2}\underline{\underline{S}}\,.$$

It should be noted, that (3.63) is an explicit time integration scheme which is why $\underline{\underline{\kappa}}$ does not depend on the stiffness matrix.

Evaluating (3.63) for $m = 0$ reveals, that not only $\hat{\underline{u}}_0$ is required for computing $\hat{\underline{u}}_1$ but also $\hat{\underline{u}}_{-1}$. While $\hat{\underline{u}}_0$ follows from the initial conditions, no obvious information is available for specifying $\hat{\underline{u}}_{-1}$. However, the latter can be approximated by a Taylor series at $t = t_0$ in direction $-h$, which yields

$$\hat{\underline{u}}_{-1} = \hat{\underline{u}}_0 - \dot{\hat{\underline{u}}}_0 h + \frac{1}{2}\ddot{\hat{\underline{u}}}_0 h^2 \tag{3.64}$$

if terms of order h^3 or higher are neglected and $\hat{\underline{u}}_{-1}$ can be estimated by means of the initial conditions for the considered problem. For the homogeneous initial conditions (3.56) and (3.57) the estimate is simply

$$\hat{\underline{u}}_{-1} = \hat{\underline{u}}_0\,.$$

For non homogeneous initial conditions, $\ddot{\hat{\underline{u}}}_0$ has to be determined from the constrained version of (3.61) at $t = 0$

$$\ddot{\hat{\underline{u}}}_0 = \underline{\underline{M}}^{-1}\left[-\underline{\underline{S}}\,\dot{\hat{\underline{u}}}_0 - \underline{\underline{K}}\,\hat{\underline{u}}_0 + \underline{q} + \hat{\underline{F}} - \underline{\Gamma}^{\mathrm{D}}\right]$$

where $\underline{\hat{u}}_0$ and $\dot{\underline{\hat{u}}}_0$ are known from the initial conditions. The result yields together with (3.64) an estimate for $\underline{\hat{u}}_{-1}$.

3.5.3 Newmark's Method

The method proposed by Newmark approximates displacements and velocities at time $t = t_{m+1}$ as follows

$$\underline{\hat{u}}_{m+1} = \underline{\hat{u}}_m + \underline{\hat{w}}_m h + \frac{h^2}{2}\left[[1-2\beta]\underline{\hat{a}}_m + \beta\,\underline{\hat{a}}_{m+1}\right] \tag{3.65}$$

$$\underline{\hat{w}}_{m+1} = \underline{\hat{w}}_m + h\left[[1-\theta]\underline{\hat{a}}_m + \theta\,\underline{\hat{a}}_{m+1}\right] \tag{3.66}$$

denoting velocities by $\underline{\hat{w}}$ and accelerations by $\underline{\hat{a}}$, i.e.,

$$\underline{\hat{w}} = \dot{\underline{\hat{u}}}\,,$$
$$\underline{\hat{a}} = \ddot{\underline{\hat{u}}}\,,$$

because it makes it easier to identify the primary unknowns. Depending on the choice of the parameters β and θ, different explicit or implicit integration methods can be obtained from (3.65) and (3.66).

Evaluating the constrained system (3.62) at $t = t_{m+1}$ and applying (3.65) as well as (3.65) yields the following system of equations with the accelerations at the nodes at time t_{m+1} as primary unknowns

$$\underline{\underline{\varkappa}}\,\underline{\hat{a}}_{m+1} = -\underline{\underline{S}}\left[\underline{\hat{v}}_m + [1-\theta]h\,\underline{\hat{a}}_m\right] - \underline{\underline{K}}\left[\underline{\hat{u}}_m + h\,\underline{\hat{w}}_m + \frac{h^2}{2}[1-2\beta]\,\underline{\hat{a}}_m\right]$$
$$+\underline{q}_{m+1} + \underline{\hat{F}}_{m+1} - \underline{\Gamma}^{\mathrm{D}}_{m+1}$$

with

$$\underline{\underline{\varkappa}} = \underline{\underline{M}} + \theta\,h\,\underline{\underline{S}} + \frac{h^2}{2}\beta\,\underline{\underline{K}}\,.$$

After computing the unknown accelerations, the unknown displacements and velocities can be determined by means of (3.65) and (3.66).

The central difference method is obtained from Newmark's method by setting $\theta = 1/2$ and $\beta = 0$ and replacing $\underline{q}_{m+1}$, $\underline{\hat{F}}_{m+1}$ and $\underline{\Gamma}^{\mathrm{D}}_{m+1}$ by $\underline{q}_m$, $\underline{\hat{F}}_m$ and $\underline{\Gamma}^{\mathrm{D}}_m$.

The method can be modified such that the nodal displacements at time t_{m+1} become the primary unknowns, which is known as modified Newmark method. Rearranging (3.65) and (3.65) for $\underline{\hat{w}}_{m+1}$ and $\underline{\hat{a}}_{m+1}$ gives

$$\underline{\hat{a}}_{m+1} = \alpha_1 \left[\underline{\hat{u}}_{m+1} - \underline{\hat{u}}_m\right] - \alpha_2 \underline{\hat{w}}_m - \alpha_3 \underline{\hat{a}}_m \tag{3.67}$$

$$\underline{\hat{w}}_{m+1} = \alpha_4 \left[\underline{\hat{u}}_{m+1} - \underline{\hat{u}}_m\right] - \alpha_5 \underline{\hat{w}}_m - \alpha_6 \underline{\hat{a}}_m \tag{3.68}$$

with

$$\alpha_1 = \frac{1}{\beta h^2}, \qquad \alpha_2 = \frac{1}{\beta h}, \qquad \alpha_3 = \frac{1-2\beta}{2\beta} h,$$

$$\alpha_4 = \frac{\theta}{\beta h}, \qquad \alpha_5 = 1 - \frac{\theta}{\beta}, \qquad \alpha_6 = \left[1 - \frac{\theta}{2\beta}\right] h.$$

For the constrained system (3.62) evaluated at at $t = t_{m+1}$, the modified Newmark method gives

$$\begin{aligned} \underline{\underline{\mathcal{K}}}\,\underline{\hat{u}}_{m+1} = & -\underline{\underline{M}}\left[\alpha_1 \underline{\hat{u}}_m + \alpha_2 \underline{\hat{w}}_m + \alpha_3 \underline{\hat{a}}_m\right] - \underline{\underline{S}}\left[\alpha_4 \underline{\hat{u}}_m - \alpha_5 \underline{\hat{w}}_m - \alpha_6 \underline{\hat{a}}_m\right] \\ & + \underline{q}_{m+1} + \underline{\hat{F}}_{m+1} - \underline{\Gamma}^{\mathrm{D}}_{m+1} \end{aligned}$$

with

$$\underline{\underline{\mathcal{K}}} = \alpha_1 \underline{\underline{M}} + \alpha_4 \underline{\underline{S}} + \underline{\underline{K}}\,.$$

After computing the unknown nodal displacements, the unknown velocities and accelerations can be determined by means of (3.67) and (3.68).

3.5.4 Stability of Time Integration Methods

The reasoning used within this section to discuss stability of time integration methods is the same as in Sect. 3.3. Two sets of initial conditions are labelled by A and B. The difference between the two sets simulates an error in the initial data, whereas a model for the evolution of this initial error in the course of time can be obtained by considering the difference between the two solutions

$$\underline{\hat{r}} = \underline{\hat{u}}^A - \underline{\hat{u}}^B\,.$$

For the constrained system (3.62) without damping the corresponding model for the evolution of $\underline{\hat{r}}$ reads

$$\underline{\underline{M}}\,\ddot{\underline{\hat{r}}} + \underline{\underline{K}}\,\underline{\hat{r}} = 0 \tag{3.69}$$

with initial conditions

$$\hat{\underline{r}}(t=0) = \hat{\underline{r}}_0 = \hat{\underline{u}}_0^A - \hat{\underline{u}}_0^B\,,$$
$$\dot{\hat{\underline{r}}}(t=0) = \dot{\hat{\underline{r}}}_0 = \dot{\hat{\underline{u}}}_0^A - \dot{\hat{\underline{u}}}_0^B\,.$$

Since (3.69) is a system of coupled differential equations, the method for investigating the stability of time integration methods laid out in Sect. 3.3 cannot be applied without diagonalising (3.69) first to obtain a system of uncoupled equations.

Diagonalisation of a matrix requires its eigenvectors, see, Sect. A.8 in the appendix. The standard ansatz

$$\hat{\underline{r}} = \hat{\underline{\eta}} \exp(I\sqrt{\omega} t)$$

with imaginary unit $I = \sqrt{-1}$ yields the eigenvalue problem

$$\underline{\underline{K}}\,\hat{\underline{\eta}} = \omega\,\underline{\underline{M}}\,\hat{\underline{\eta}} \tag{3.70}$$

which can be cast into the standard form by multiplying both sides by the inverse mass matrix. The solution of (3.70) are the N_d eigenvalues ω_i and their corresponding eigenvectors $\hat{\underline{\eta}}_i$ of the system, where N_d indicates its number of degrees of freedom.

More information about the eigenvectors can be obtained by considering (3.70) and its transpose for specific eigenvectors and eigenvalues, i.e.,

$$\underline{\underline{K}}\,\hat{\underline{\eta}}_i = \omega_i\,\underline{\underline{M}}\,\hat{\underline{\eta}}_i \qquad \hat{\underline{\eta}}_k^{\mathrm{T}}\,\underline{\underline{K}} = \omega_k\,\hat{\underline{\eta}}_k^{\mathrm{T}}\,\underline{\underline{M}}$$

and multiplying these equations with particular eigenvectors, such that

$$\hat{\underline{\eta}}_k^{\mathrm{T}}\,\underline{\underline{K}}\,\hat{\underline{\eta}}_i = \omega_i\,\underline{\underline{M}}\,\hat{\underline{\eta}}_i \qquad \hat{\underline{\eta}}_k^{\mathrm{T}}\,\underline{\underline{K}}\,\hat{\underline{\eta}}_i = \omega_k\,\hat{\underline{\eta}}_k^{\mathrm{T}}\,\underline{\underline{M}}\,\hat{\underline{\eta}}_i$$

is obtained. The difference between the results reads

$$[\omega_i - \omega_k]\,\hat{\underline{\eta}}_k^{\mathrm{T}}\,\underline{\underline{M}}\,\hat{\underline{\eta}}_i = 0 \tag{3.71}$$

which leads to the following conclusion. For $i = k$, (3.71) is automatically fulfilled because the difference of the eigenvalues vanishes. For $i \neq k$, however, the remaining factor on the left hand side of (3.71) must vanish, which yields

$$\hat{\underline{\eta}}_k^{\mathrm{T}}\,\underline{\underline{M}}\,\hat{\underline{\eta}}_i = \delta_{ki} \tag{3.72}$$

provided, that eigenvectors are properly normalised. This result implies

$$\hat{\underline{\eta}}_k^{\mathrm{T}}\,\underline{\underline{K}}\,\hat{\underline{\eta}}_i = \begin{cases} \omega_k & i = k \\ 0 & i \neq k \end{cases}\,. \tag{3.73}$$

The solution of (3.69) can be written in general as the sum of all eigenvectors and corresponding time-dependent coefficients as follows

$$\underline{\hat{r}} = \sum_{i=1}^{N_d} g_i \, \underline{\hat{\eta}}_i \, . \tag{3.74}$$

Introducing (3.74) into (3.69), multiplying from the left by $\underline{\hat{\eta}}_k^{\mathrm{T}}$, and taking into account (3.72) as well as (3.73) yields the following system of uncoupled equations for the time-dependent coefficients $g_i, i = 1, \ldots, N_d$

$$\ddot{g}_i + \omega_i \, g_i = 0$$

Every equation can now be discussed separately in the same way as discussed in Sect. 3.3. In the following, the index referring to a particular equation of the uncoupled system is dropped for better readability and

$$\ddot{g} + \omega g = 0 \tag{3.75}$$

is considered for discussing the main aspects of the approach.

Exemplary, the integration scheme based on the difference quotient (3.20) is examined. Evaluating (3.75) at time $t = t_m$ and applying (3.20) gives after regrouping

$$g_{m+1} - g_m[2 - \omega h^2] + g_{m-1} \, . \tag{3.76}$$

and the quadratic equation

$$Z^2 - Z[2 - \omega h^2] + Z = 0 \tag{3.77}$$

is obtained by introducing the ansatz

$$g_m = Z^m \, g_0$$

into (3.76) and multiplying the result with $Z^{-(m+1)}$. The solution of equation (3.77) reads

$$Z_{1,2} = \frac{2 - \omega h^2}{2} \pm \sqrt{\frac{[2 - \omega h^2]^2}{4} - 1} \, .$$

For the time integration to be stable, $|Z| < 1$ must hold which reveals after careful examination

$$h < \frac{2}{\sqrt{|\omega|}} \, . \tag{3.78}$$

Because it is supposed, that $\underline{\underline{K}}$ and $\underline{\underline{M}}$ are positive definite, the absolute value signs in the above equation can actually be omitted.

The value for h in (3.78) decreases if ω increases. Since a system like (3.62) has N_d eigenvalues, the largest eigenvalue determines the upper bound for h. For instance, in the case of axial deformation, the largest eigenvalue of a linear finite element e is given by

$$\omega_e^{\max} = \frac{8\lambda}{\rho L_e^2}$$

with stiffness λ, density ρ, and element length L_e, which indicates that the length of the smallest element determines the upper bound for h to ensure stability of the time integration method.

Remark 3.4 (*Damping*) The method laid out in this section applies as well for damped systems, provided, that eigenvalues and eigenvectors for the considered system with and without damping coincide. The latter is true, if Rayleigh damping is an appropriate model, which means, that the damping matrix can be written, for instance, as

$$\underline{\underline{\boldsymbol{S}}} = \gamma \underline{\underline{\boldsymbol{M}}} .$$

Applying (3.20) together with the central difference quotient, see (3.19), for the system with damping, stability can be studied analogously to the case without damping and

$$h < \frac{2}{\sqrt{|\omega|}} \left[\sqrt{\gamma^2 + 1} - \gamma \right]$$

is obtained eventually.

3.6 Bibliographical Remarks

The reader is referred to [1] for further reading on initial value problems and differential equations in general. Regarding time integration methods and, in particlar, Runge-Kutta methods, a large and constantly growing body of literature exists which makes it difficult to recommend a particular text book. One of the author's personal favorits is [2] but other equally valuable sources are available.

A similar situation is encountrered for structural dynamics. Therefore, [3, 4] are only two of many recommendable books on the subject.

Initial boundary value problems are problems with space-time coupling. As already mentioned, decoupling can be achieved by assuming time derivatives to be constant within time steps. Under these circumstances, the methods of Functional Analysis, outlined in Chap. C, apply without restrictions to the spatial part. Otherwise, more elaborated methods are required for discussing uniqueness and stability of solutions. A general framework for initial boundary value problems is given in [5].

References

1. T. Myint-U, L. Debnath, *Linear Partial Differential Equations for Scientists and Engineers* (Birkhäuser Boston, 2007)
2. S. Nakamura, *Applied Numerical Methods with Software*. Prentice-Hall international editions (Prentice Hall, 1991)
3. A. Shabana, *Theory of Vibration: An Introduction* (Mechanical Engineering Series (Springer, New York, 2011)
4. R. Craig, A. Kurdila, *Fundamentals of Structural Dynamics* (Wiley, 2011)
5. K. Surana, J. Reddy, *The Finite Element Method for Initial Value Problems: Mathematics and Computations* (CRC Press, 2017)

Chapter 4
Non-linear Boundary Value Problems

Abstract After discussing possible sources of non-linearities, the subject is discussed for the case of non-linear constitutive equations and spatially one-dimensional problems. Developing FEM for non-linear boundary value problems leads eventually to systems of non-linear algebraic equations and the use of Newton's method for solving such systems is demonstrated. The commonly applied approach for controlling convergence of Newton's method by employing a time or pseudo time incrementation procedure, respectively, with nested iteration loop is laid out. Numerical integration employing Gauss integration is discussed together with its benefits regarding the separation of spatial discretisation and material routines.

4.1 Non-linear Poisson Equation in $\mathbb{R}$

4.1.1 Strong Form

There are several possible sources of non-linearities with different order of complexity. The most simple case are non-linear constitutive equations, followed by non-linear boundary conditions, and non-linear changes of the domain Ω. Mechanical contact is a standard example for non-linear boundary conditions and large deformations in Continuum Mechanics obviously imply significant changes regarding the domain Ω.

Here, only non-linear constitutive equations are considered. However, many of the concepts discussed in this chapter apply as well for more complex situations. The main aspects are illustrated for axial deformation, see Example 2.1. Equilibrium conditions (2.1) and kinematics (2.2) remain unchanged but the relation between axial force P and strain ϵ can be given in terms of an algebraic equation or even as differential equation.

Therefore, the strong form of the boundary value problem considered in the sequel is defined be the following set of equations to be satisfied for $x \in \Omega = (0, l)$

U. Mühlich, *Enhanced Introduction to Finite Elements for Engineers*, Solid Mechanics and Its Applications 268, https://doi.org/10.1007/978-3-031-30422-4_4

$$P' + n = 0\,, \tag{2.1 revisited}$$

$$\epsilon = u'\,, \tag{2.2 revisited}$$

$$P \leftrightarrow \epsilon\,. \tag{4.1}$$

In addition, the boundary conditions

$$u(0) = 0 \tag{4.2}$$

$$P(l) = P_l \tag{4.3}$$

are considered here exemplary.

4.1.2 Variational and Weak Form

Analogously to Sect. 2.1.3, (2.1) is multiplied by a test function v and the result is integrated with respect to the domain Ω

$$\int_\Omega P' v \,\mathrm{d}x + \int_\Omega n\, v \mathrm{d}x = 0\,.$$

Applying integration by parts to the first term of the left hand side gives

$$\int_\Omega P\, v' \,\mathrm{d}x - \int_\Omega n\, v \mathrm{d}x + P(0)v(0) - P(l)v(l) = 0\,.$$

Considering only test functions which vanish in the presence of essential boundary conditions and taking into account (4.3), the variational form

$$\mathcal{W} = \int_\Omega P\, v' \,\mathrm{d}x - \int_\Omega n\, v \mathrm{d}x - P_l v(l) = 0 \tag{4.4}$$

$$u(0) = 0\,, \tag{4.5}$$

is derived. The corresponding weak form can be obtained by interpreting derivatives as weak derivatives. Although (4.4) is of form

$$a(u, v) - b(v) = 0\,,$$

the Lax-Milgram lemma does obviously not apply because, since P is in general a non-linear function of u, $a(u, v)$ is not bilinear, see Appendix C.

There are at least two ways to proceed. The variational form (4.4) can be linearised, using, for instance, the Gateaux derivative, to define an iterative procedure which consists in solving a sequence of linear variational problems using FEM. Alternatively, the FEM solution scheme can be developed directly for the non-linear variational problem which leads eventually to a system of non-linear equations. The latter is solved by solving a sequence of systems of linear equations. Here, the second option is preferred.

4.1.3 Galerkin FEM

Linear Lagrange elements are used to illustrate main aspects of FEM solution schemes for non-linear boundary value problems. Therefore, the FEM discretisation scheme discussed in the following is in large parts identical to the procedure for linear boundary value problems discussed in Sect. 2.1.6. Again, piecewise integration is applied, hence

$$\mathcal{W} = \sum_{e=1}^{N_e} \mathcal{W}_e + P(l)v(l) = 0 \tag{4.6}$$

with

$$\mathcal{W}_e = \int\limits_{\Omega_e} P\, v\, \mathrm{d}x - \int\limits_{\Omega_e} n\, v\, \mathrm{d}x\,.$$

A reference domain $\Omega_\square$ and corresponding mappings χ_e are defined. The coordinates x and ξ are used to address points of the Ω and $\Omega_\square$. Only main results of Sect. 2.1.7 are reiterated for the benefit of the reader.

Recall, that the trial functions defined for the reference domain $\Omega_\square$ and their derivatives can be written concisely as

$$u_\square(\xi) = \underline{\boldsymbol{N}}^{\mathrm{T}}\, \underline{\hat{\boldsymbol{u}}}_\square \tag{2.37 revisited}$$

$$u'_\square(\xi) = \underline{\boldsymbol{B}}^{\mathrm{T}}\, \underline{\hat{\boldsymbol{u}}}_\square\, j_e^{-1}\,. \tag{2.38 revisited}$$

The test functions and their derivatives take a similar form

$$v_\square(\xi) = \underline{\boldsymbol{N}}^{\mathrm{T}}\, \underline{\hat{\boldsymbol{v}}}_\square \tag{2.39 revisited}$$

$$v'_{\square}(\xi) = \underline{\boldsymbol{B}}^{\mathrm{T}}\, \underline{\hat{\boldsymbol{v}}}_{\square}\, j_e^{-1} \tag{2.40 revisited}$$

with

$$\underline{\boldsymbol{N}}^{\mathrm{T}} = \left[\tfrac{1}{2}(1-\xi)\;\; \tfrac{1}{2}(1+\xi)\right] \qquad\qquad \underline{\boldsymbol{B}}^{\mathrm{T}} = \left[-\tfrac{1}{2}\;\; \tfrac{1}{2}\right]$$

and

$$\underline{\hat{\boldsymbol{u}}}_{\square}^{\mathrm{T}} = \left[u_0\; u_1\right], \qquad\qquad \underline{\hat{\boldsymbol{v}}}_{\square}^{\mathrm{T}} = \left[v_0\; v_1\right].$$

The mapping χ_e relates the reference domain with a particular domain Ω_e. Its Jacobian is denoted by j_e.

The contribution of a domain Ω_e expressed by means of trial and test functions referring to $\Omega_{\square}$ eventually reads

$$\mathcal{W}_{e_{\square}} = \underline{\hat{\boldsymbol{v}}}_{\square}^{\mathrm{T}} \left[\underline{\boldsymbol{H}}_e - \underline{\boldsymbol{q}}_e\right] \tag{4.7}$$

with

$$\underline{\boldsymbol{H}}_e = \int_{-1}^{+1} \underline{\boldsymbol{B}}\, P\, \frac{1}{j_e}\, \mathrm{d}\xi \tag{4.8}$$

and the element load vector

$$\underline{\boldsymbol{q}}_e = \int_{-1}^{+1} n\, \underline{\boldsymbol{N}}\, j_e\, \mathrm{d}\xi\,. \tag{2.45 revisited}$$

The contribution of an element e to the weak form, $\mathcal{W}_e$, is obtained from (4.7) by applying (2.41), which gives

$$\mathcal{W}_e = \underline{\hat{\boldsymbol{v}}}^{\mathrm{T}} \left[\underline{\underline{\boldsymbol{A}}}_e^{\mathrm{T}}\, \underline{\boldsymbol{H}}_e - \underline{\underline{\boldsymbol{A}}}_e^{\mathrm{T}}\, \underline{\boldsymbol{q}}_e\right] \tag{4.9}$$

with gathering matrices $\underline{\underline{\boldsymbol{A}}}_e$. The discretised weak form is finally obtained by introducing (4.9) into (4.6). The final result is

$$\mathcal{W}_h = \underline{\hat{\boldsymbol{v}}}^{\mathrm{T}} \left[\underline{\boldsymbol{H}}\,\underline{\hat{\boldsymbol{u}}} - \underline{\boldsymbol{q}}\,\underline{\hat{\boldsymbol{u}}} - \underline{\hat{\boldsymbol{F}}}\right] = 0 \tag{4.10}$$

with

$$\underline{\boldsymbol{H}} = \sum_{e=1}^{N} \underline{\underline{\boldsymbol{A}}}_e^{\mathrm{T}}\, \underline{\boldsymbol{H}}_e \tag{4.11}$$

and

$$\underline{q} = \sum_{e=1}^{N} \underline{\underline{A}}_e^{\mathrm{T}} \underline{q}_e \,. \tag{4.12}$$

Regarding the definition of $\underline{\hat{F}}$, see the discussion below Eq. (2.26).

Because of the arbitrariness of $\underline{\hat{v}}$, (4.10) can only be fulfilled in general if the bracket vanishes, which yields the unconstrained system of equations

$$\underline{G} = \underline{H} - \underline{q} - \underline{\hat{F}} = \underline{0} \,. \tag{4.13}$$

The corresponding constrained system is obtained by incorporating essential boundary conditions. Because P in (4.8) depends in general non-linearly on u', (4.13) is a system of non-linear equations. One of the most widely used methods for solving non-linear equations is Newton's method to be discussed next.

4.2 Newton's Method for Solving Non-linear Equations

Given a non-linear equation $f(x) = 0$ for $x \in [0, 2]$, where $f(x)$ can be a highly non-linear function, for instance, $f(x) = \tanh(2x - 1))/8 - \sin x - \cos x + x^2$. The key idea behind Newton's method is to find the solution of such a non-linear equation by solving a sequence of linear equations.

Starting with some initial guess x_k, Taylor series expansion of the non-linear equation $f(x) = 0$ at the initial guess reads

$$f(x_k + \blacktriangle x) = f(x_k) + f'(x_k)\blacktriangle x + O(\blacktriangle x^2) = 0,$$

which, if all terms of order $\blacktriangle x^2$ or higher are neglected, gives a linear equation

$$f(x_k + \blacktriangle x) = f(x_k) + f'(x_k)\blacktriangle x = 0$$

which has the solution

$$\blacktriangle x = -\frac{f(x_k)}{f'(x_k)} \,.$$

The question arises if $x_{k+1} = x_k + \blacktriangle x$ fulfils the non-linear equation $f(x_{k+1}) = 0.$, which is usually not the case and the algorithm proceeds now from x_{k+1} and so forth, until the non-linear equation is fulfilled with desired accuracy. Provided a suitable initial guess, convergence of Newton's method is quadratic. For a pseudocode version of Newton's method see Algorithm 2.

Algorithm 2 Newton's method for solving $f(x) = 0$

Require: initial guess x_0, precision ε, e.g., $\varepsilon = 10^{-6}$, k_{max}
Ensure: x_f provided convergence, conv

$x \leftarrow 0$
$k \leftarrow 0$

while $(k < k_{max}$ **and** $|f(x)| > \varepsilon)$ **do**

$\blacktriangle x = -\frac{f(x)}{f'(x)}$

$x \leftarrow x + \blacktriangle x$
$k \leftarrow k + 1$

end while
if $(k \geq k_{max}$ **and** $|f(x_f)| > \varepsilon)$ **then**
conv $\leftarrow$ false ▷ no convergence
else
$x_f \leftarrow x$
conv $\leftarrow$ true
end if

Newton's method also known as Newton-Raphson method, can be extended to systems of non-linear equations. However, since the convergence of the method depends heavily on a suitable initial guess, its direct application to the constrained version of (4.13) is not recommendable.

4.3 Solving the System of Non-linear FEM Equations

Possible convergence problems of Newton's method can be reduced either by using methods for estimating suitable start values or by increasing loads gradually starting from zero instead of applying them at once. Here, the latter option is discussed as it is the preferred choice in most FEM software. Loading is parametrised by a load parameter or pseudo time, which can also be an actual time. Therefore, we refer to this parameter simply as time in the following.

It is assumed, that the solution is known at a time $t = t_m$. The unknowns are the nodal displacements at the end of a time increment $[t_m, t_{m+1} = t_m + \Delta t]$, $\underline{\hat{\boldsymbol{u}}}_{m+1}$. The latter can be expressed as

$$\underline{\hat{\boldsymbol{u}}}_{m+1} = \underline{\hat{\boldsymbol{u}}}_m + \Delta\underline{\hat{\boldsymbol{u}}},$$

to elucidate, that the unknowns for a time increment are actually the changes in nodal displacements $\Delta\underline{\hat{\boldsymbol{u}}}$. Therefore, the constrained version of (4.13) is actually a system of non-linear equations for $\Delta\underline{\hat{\boldsymbol{u}}}$, more specifically, the problem to be solved reads

$$\underline{G}(\hat{\underline{u}}_m + \Delta\hat{\underline{u}}) = \underline{0}\,. \tag{4.14}$$

Starting with an initial guess $\Delta\hat{\underline{u}}_k$, Taylor expansion of the left hand side of (4.14) at $\Delta\hat{\underline{u}}_k$ reads

$$\underline{G}(\hat{\underline{u}}_m + \Delta\hat{\underline{u}}_k + \blacktriangle\Delta\hat{\underline{u}}) = \underline{G}(\hat{\underline{u}}_m + \Delta\hat{\underline{u}}_k) + \left.\frac{\mathrm{d}\underline{G}}{\mathrm{d}\hat{\underline{u}}}\right|_{\hat{\underline{u}}_m+\Delta\hat{\underline{u}}_k} \blacktriangle\Delta\hat{\underline{u}}$$

if all higher order terms with respect to $\blacktriangle\Delta\hat{\underline{u}}$ are neglected. In the following, the more condensed notation

$$\underline{G}(\hat{\underline{u}}_{m:k} + \blacktriangle\Delta\hat{\underline{u}}) = \underline{G}_{m:k} + \left.\frac{\mathrm{d}\underline{G}}{\mathrm{d}\hat{\underline{u}}}\right|_{m:k} \blacktriangle\Delta\hat{\underline{u}}$$

is used for the sake of readability. Plugging the truncated Taylor expansion into (4.14) leads to the following system of linear equations

$$\underline{G}_{m:k} + \left.\frac{\mathrm{d}\underline{G}}{\mathrm{d}\hat{\underline{u}}}\right|_{m:k} \blacktriangle\Delta\hat{\underline{u}} = \underline{0} \tag{4.15}$$

for $\blacktriangle\Delta\hat{\underline{u}}$, which forms the heart of Newton's method for solving (4.14).

To apply (4.15), the derivative of $\underline{G}$ with respect to the nodal displacements has to be specified. The individual terms in (4.13) are considered separately, starting with

$$\frac{\mathrm{d}\underline{H}}{\mathrm{d}\hat{\underline{u}}} = \frac{\mathrm{d}}{\mathrm{d}\hat{\underline{u}}}\sum_{e=1}^{N_e} \underline{\underline{A}}_e^{\mathrm{T}}\,\underline{H}_e = \sum_{e=1}^{N_e} \underline{\underline{A}}_e^{\mathrm{T}} \frac{\mathrm{d}}{\mathrm{d}\hat{\underline{u}}}\underline{H}_e\,.$$

According to (4.8), $\underline{H}_e$ depends on P which in turn depends on ϵ, and the latter depends on u'. Therefore

$$\frac{\mathrm{d}}{\mathrm{d}\hat{\underline{u}}}\underline{H}_e = \int_{-1}^{+1} \underline{B}\frac{\mathrm{d}}{\mathrm{d}\hat{\underline{u}}}P\, j_e^{-1}\mathrm{d}x = \int_{-1}^{+1} \underline{B}\frac{\mathrm{d}}{\mathrm{d}\hat{\underline{u}}}P(\epsilon)\, j_e^{-1}\mathrm{d}x\,. \tag{4.16}$$

Taking into account, that ϵ for a domain e is given by $\epsilon = \underline{B}^{\mathrm{T}}\,\underline{\underline{A}}_e\,\hat{\underline{u}}$ and applying chain rule yields eventually

$$\frac{\mathrm{d}}{\mathrm{d}\hat{\underline{u}}}\underline{H}_e = \underline{\underline{K}}_e\,\underline{\underline{A}}_e \tag{4.17}$$

with element stiffness matrix $\underline{\underline{K}}_e$ and material tangent C given by

$$\underline{\underline{K}}_e = \int_{-1}^{+1} \underline{B}\, C\, \underline{B}^{\mathrm{T}}\, j_e^{-1} \mathrm{d}x \qquad \text{and} \qquad C = \frac{\mathrm{d}}{\mathrm{d}\epsilon} P(\epsilon)\,. \tag{4.18}$$

As expected, the element the stiffness matrix for the linear case is obtained if C in $\underline{\underline{K}}_e$ is replaced by λ, see (2.44).

Using, (4.17), (4.16) can be written simply as

$$\frac{\mathrm{d}\underline{H}}{\mathrm{d}\underline{\hat{u}}} = \sum_{e=1}^{N_e} \underline{\underline{A}}_e^{\mathrm{T}} \underline{\underline{K}}_e \underline{\underline{A}}_e = \underline{\underline{K}}\,.$$

If n and P_l in (4.4) depend on u, the derivatives of $\underline{\hat{F}}$ and $\underline{q}$ with respect to $\underline{\hat{u}}$ are required as well. However, here only conservative loads are considered for the sake of simplicity. In this case, only the requested load level at the end of the increment matters and (4.15) reads

$$\underline{\underline{K}}_{m:k} \blacktriangle\Delta\underline{\hat{u}} = -\underline{H}_{m:k} + \underline{q}_{m+1} + \underline{\hat{F}}_{m+1}\,. \tag{4.19}$$

A closer look at (4.14), (4.15), (4.19) and (4.18) reveals a number of observation that are helpful regarding the implementation of the method:

1. A suitable and obvious initial guess for (4.19) is $\Delta\underline{\hat{u}} = \underline{0}$. If there is no convergence for a time increment of length Δt, starting a new attempt with a smaller Δt is a plausible strategy provided, of course, a solution exists.
2. The structure of $\underline{\underline{K}}_e$ for Newton's method is identical with that of the linear element stiffness matrix. It is worth noting, that information about spatial discretisation and constitutive behaviour is nicely separated, i.e., $\underline{B}$ contains all necessary information about spatial discretisation and interpolation order, whereas C provides the information regarding the material behaviour. Solving the integral in (4.18) numerically provides a way for implementing material routines separately, see Sect. 4.4.

It can readily be seen, that the implementation of a FEM solution scheme for solving non-linear boundary value problems is much more complex than in the case of linear boundary value problems. Not only a time or pseudo time discretisation is required but for every time increment a system of non-linear equations has to be solved using an iterative procedure, which may or may not converge. This complexity also places higher demands on the users of FEM software in terms of professional skills and experience level. For the pseudocode of a simplified non-linear FEM solution procedure, see Algorithm 3.

Example 4.1 (*Derivative of P with respect to $\underline{\hat{u}}$*) Consider the constitutive equation $P = \alpha\epsilon^m$, which for the domain Ω_e takes the form

$$P = \alpha \left[\underline{B}^{\mathrm{T}} \underline{\underline{A}}_e \underline{\hat{u}}\right]^m .$$

Differentiation of P with respect to $\underline{\hat{u}}$, applying chain rule, gives

$$\frac{\mathrm{d}P}{\mathrm{d}\underline{\hat{u}}} = \alpha[m-1]\epsilon^{m-1}\underline{\underline{B}}^{\mathrm{T}}\underline{\underline{A}}_e$$

and the reader is invited to perform the calculation using index notation taking into account that for the unconstrained system

$$\frac{\mathrm{d}P}{\mathrm{d}\underline{\hat{u}}}^{\mathrm{T}} = \left[\frac{\partial P}{\partial u_1}\ \frac{\partial P}{\partial u_2} \cdots \frac{\partial P}{\partial u_{N_{\mathrm{dofs}}}}\right]$$

holds.

4.4 Gauss-Legendre Integration

The numerical integration method known as Gauss-Legendre method is the preferred choice in most FEM software for computing the integrals of element stiffness matrices and similar tasks. Numerical integration procedures possess the general form

$$\int_{-1}^{+1} f(\xi)\,\mathrm{d}\xi \approx \sum_{i=1}^{N} w_i\, f(\xi_i) \tag{4.20}$$

for an interval $\xi \in [-1, 1]$ with N nodes at which the given continuous function $f(\xi)$ is evaluated and multiplied with corresponding weights w_i.

The right hand side of (4.20) reveals, that the result depends on the positions ξ_i. Using equidistant positions seems to be a natural choice but, as discussed below, it is not necessarily the best one.

If $f(\xi)$ is a polynomial of degree M, see, (2.79) together with the corresponding text, it can be written as the sum of powers of ξ multiplied by corresponding coefficients, i.e.,

$$f(\xi) = c_0 + c_1\xi + c_2\xi^2 + \cdots + c_M\xi^M = \sum_{i=0}^{M} c_i\, \xi^i\,.$$

Each term of the polynomial can be integrated exactly and the result is a sum with $M + 1$ terms and it can be shown, that (4.20) is exact at least for polynomials of degree less than N provided a suitable choice of the weights w_i.

However, weights and positions in (4.20) can be chosen such that (4.20) is even exact for polynomials of degree less that $2N$. Showing this, requires some knowledge about Legendre polynomials, called $\mathcal{P}_i(\xi)$ in the following.

Example 4.2 (*Legendre polynomials*) The first four Legendre polynomials are [1]

$$\mathcal{P}_0 = 1\,, \qquad \mathcal{P}_1 = \xi\,, \qquad \mathcal{P}_2 = \frac{1}{2}\left[3\xi^2 - 1\right], \qquad \mathcal{P}_3 = \frac{1}{2}\left[5\xi^3 - 3x\right].$$

Please note, that indexing of the Legendre polynomials starts with zero.

Legendre polynomials possess a number of nice properties, where especially two of them are important here. Firstly,

$$\int_{-1}^{+1} P^i(\xi)\,\mathcal{P}_N(\xi)\,\mathrm{d}\xi = 0 \quad \text{if} \quad i < N-1 \tag{4.21}$$

holds and, secondly, the Legendre polynomial $\mathcal{P}_n$ has exactly n real roots $\hat{\xi}_j$, which all lie in the interval $(-1, +1)$.

Furthermore, according to the Euclidean division theorem for polynomials [2], a polynomial of degree m can be written as

$$P^m(\xi) = P^i(\xi)\,P^N(\xi) + P_\mathrm{r}^k(\xi) \tag{4.22}$$

with $k < i$, where P^N and $P_\mathrm{r}^k(\xi)$ are called divisor and remainder, respectively. Since the degree of the product of two polynomials equals the sum of their degrees, $i = N - 1$ must hold.

Considering $m = 2N - 1$ and setting $P^N = \mathcal{P}_{N+1}$, (4.22) reads

$$P^{2N-1}(\xi) = P^i(\xi)\,\mathcal{P}_{N+1}(\xi) + P_\mathrm{r}^k(\xi) \tag{4.23}$$

with $i = N - 1$ and $k < N - 1$. Integration yields

$$\int_{-1}^{+1} P^{2N-1}(\xi)\,\mathrm{d}\xi = \int_{-1}^{+1} P^i(\xi)\,\mathcal{P}_{N+1}(\xi)\,\mathrm{d}\xi + \int_{-1}^{+1} P_\mathrm{r}^k(\xi)\,\mathrm{d}\xi = \int_{-1}^{+1} P_\mathrm{r}^k(\xi)\,\mathrm{d}\xi$$

due to (4.21). Since $k < N - 1$, exact integration can be achieved by

$$\int_{-1}^{+1} P^{2N-1}(\xi)\,\mathrm{d}\xi = \sum_{j=1}^{N} w_j\, P_\mathrm{r}^k(\xi_j)\,.$$

On the other hand,

$$\sum_{j=1}^{N} P^{2N-1}(\xi_j) = \sum_{j=1}^{N} P^i(\xi_j)\,\mathcal{P}_{N+1}(\xi) + \sum_{j=1}^{N} P_\mathrm{r}^k(\xi_j)$$

Table 4.1 Roots $\hat{\xi}_j$ and weights w_j for Gauss-Legendre integration up to order three. Regarding the corresponding Legendre polynomials, see Example 4.2

n	$\hat{\xi}_j$	w_j
1	$\hat{\xi}_1 = 0$	$w_1 = 2$
2	$\hat{\xi}_1 = -\frac{1}{\sqrt{3}}, \hat{\xi}_2 = \frac{1}{\sqrt{3}}$	$w_1 = 1, w_2 = 1$
3	$\hat{\xi}_1 = 0, \hat{\xi}_2 = -\sqrt{\frac{3}{5}}, \hat{\xi}_3 = \sqrt{\frac{3}{5}}$	$w_1 = \frac{5}{9}, w_2 = \frac{8}{9}, w_3 = \frac{8}{9}$

holds. If the coordinates ξ_j are chosen to be the N roots of $\mathcal{P}_{N+1}$, i.e., the positions $\hat{\xi}_j$ at which $\mathcal{P}_{N+1}$ vanishes,

$$\sum_{j=1}^{N} P^{2N-1}(\hat{\xi}_j) = \sum_{j=1}^{N} P_{\mathrm{r}}^{k}(\hat{\xi}_j)$$

is obtained from which

$$\int_{-1}^{+1} P^{2N-1}(\xi)\, \mathrm{d}\xi = \sum_{j=1}^{N} w_j\, P^{2N-1}(\hat{\xi}_j)$$

can be deduced. The argument starting with (4.23) applies analogously to cases with $m < 2N - 1$. Therefore, the general Gauss-Legendre integration formula

$$\int_{-1}^{+1} f(\xi)\, \mathrm{d}\xi \approx \sum_{j=1}^{N} w_j\, f(\hat{\xi}_j) \quad (4.24)$$

gives the exact result if f is a polynomial of degree less or equal to $2N - 1$. For $\mathcal{P}_1, \mathcal{P}_1$ and $\mathcal{P}_3$ given in Example 4.2, the roots, which in the context of FEM are often simply called Gauss points, can be read off easily. The corresponding weights w_j are given by [3]

$$w_j = \int_{-1}^{+1} \frac{\mathcal{P}_n(\xi)}{[\xi - \hat{\xi}_j]\mathcal{P}'_n(\hat{\xi}_j)}\, \mathrm{d}\xi = \frac{2}{[1 - \hat{\xi}_j^2][\mathcal{P}'_n(\hat{\xi}_j)]^2}\,.$$

The results are summarised in Table 4.1. For higher degrees, the reader is referred to [3] or standard text books. The method extends to higher dimensions as well.

4.5 Implementation Aspects

Gauss-Legendre integration is used in most FEM software to compute $\underline{\underline{K}}_e$ and $\underline{H}_e$, see (4.18) and (4.8). For an integration order N_G, $\underline{\underline{K}}_e$ and are computed as follows

$$\underline{\underline{K}}_e \approx \sum_{j=1}^{N_G} \underline{B}(\hat{\xi}_j)\, C(\hat{\xi}_j) \underline{B}^{\mathrm{T}}(\hat{\xi}_j)\, j_e^{-1}(\hat{\xi}_j)\,,$$

$$\underline{H}_e \approx \sum_{j=1}^{N_G} \underline{B}(\hat{\xi}_j)\, P(\hat{\xi}_j)\, j_e^{-1}(\hat{\xi}_j)\,,$$

which allows for separating interpolation order from constitutive behaviour. More specifically, a finite element implementation for a particular interpolation order can be combined with different constitutive laws provided, that corresponding material routines exist which can be called during the numerical integration. These routines must return C and P at the integration points according to the current state of the Newton iteration.

It is worth noting, that only the values of the essential unknowns, such as u, are computed at the nodes of a finite element mesh. Derived quantities like P are originally computed at the integration points. FEM software solutions commonly provide values for these quantities as well at the nodes. This, however, requires some averaging procedure. Therefore, nodal values of derived quantities at mesh nodes are less accurate than corresponding values at integration points.

Implementing material routines can be a delicate issue if the constitutive behaviour involves differential equations which have to be integrated with respect to time or pseudo time. In order to ensure quadratic convergence of the Newton iteration, the material tangent C has to be consistent with the integration procedure. A typical example is rate independent plasticity.

Since, the theory of plasticity is beyond the scope of this book, the problem is illustrated by means of an incremental form of non-linear elasticity. Considering the constitutive law

$$P = \alpha \epsilon^{\beta}$$

with material constants α and β, which can be parametrised by a time or pseudo time variable t. The change of P with respect to t is controlled by the derivative

$$\frac{\mathrm{d}P}{\mathrm{d}t} = \frac{\mathrm{d}P}{\mathrm{d}\epsilon}\frac{\mathrm{d}\epsilon}{\mathrm{d}t} = \alpha[\beta - 1]\epsilon^{\beta-1}\frac{\mathrm{d}\epsilon}{\mathrm{d}t} = \alpha[\beta - 1]\epsilon^{\beta-1}\dot{\epsilon}\,.$$

Approximating $\dot{\varepsilon}$ in the k-th iteration step by

$$\dot{\varepsilon}_k = \frac{\Delta\epsilon_k}{\Delta t}\,,$$

the task of integrating the constitutive equation in the k-th iteration step can be expressed in the language established in Sect. 3.2 as follows

$$P_{m:k} = P_m + \int_{t_m}^{t_{m+1}} g \, \mathrm{d}t$$

with

$$g = \alpha[\beta - 1]\epsilon^{\beta-1}\Delta\epsilon_k \, .$$

Performing the integration using Euler's explicit method yields

$$P_{m:k} = P_m + \alpha[\beta - 1]\,[\epsilon_m]^{\beta-1}\,\Delta\epsilon_k \, , \tag{4.25}$$

whereas

$$P_{m:k} = P_m + \alpha[\beta - 1]\,[\epsilon_m + \Delta\epsilon_k]^{\beta-1}\,\Delta\epsilon_k \, , \tag{4.26}$$

is obtained for Eulers's implicit method. The consistent tangent of the Newton iteration is defined by

$$C_{m:k} = \frac{\mathrm{d}P_{m:k}}{\mathrm{d}\epsilon_{m:k}} = \frac{\mathrm{d}P_{m:k}}{\mathrm{d}\Delta\epsilon_k} \tag{4.27}$$

because of $\epsilon_{m:k} = \epsilon_m + \Delta\epsilon_k$. The corresponding tangents read

$$C_{m:k} = \alpha[\beta - 1]\begin{cases} [\epsilon_m]^{\beta-1} & \text{Euler explicit} \\ [\epsilon_{m:k} + \Delta\epsilon_k]^{\beta-1} + [\beta - 1]\,[\epsilon_m + \Delta\epsilon_k]^{\beta-2}\,\Delta\epsilon_k & \text{Euler implicit} \end{cases}$$

and incorrect implementation of the consistent tangent can provoke serious convergence problems, especially for complex material models and larger systems.

4.6 Bibliographical Remarks

Compared to linear boundary value problems, applying Finite Element Methods to non-linear boundary value problems requires significantly more knowledge not only in terms of numerical methods but as well about the boundary value problem as such.

Non-linear Continuum Mechanics, for instance, is a broad and complex field, including various sub-disciplines. Regarding Non-linear Elasticity, Theory of Plasticity, and Non-linear Fracture Mechanics, the reader is referred to [4–6], respectively.

The complexity of a subject increases significantly if large deformations and / or large rotations have to be taken into account. To the same extent increases the

Algorithm 3 Pseudocode of a simplified non-linear FEM implementation

Require: end time t_{E}, time increment Δt, precision ε, e.g., $\varepsilon = 10^{-6}$, $k_{\max}$, n, P_l
Ensure: $\underline{\hat{u}}$ at $t = t_{\mathrm{E}}$ provided convergence, conv

$t \leftarrow 0$
$\underline{\hat{u}}_{m+1} \leftarrow \underline{0}$

conv $\leftarrow$ true

while ($t < t_{\mathrm{E}}$) **and** conv= true) **do** ▷ TIME LOOP

  $\bar{n} \leftarrow n\,(t + \Delta t)$, $\bar{P}_l \leftarrow P_l(t + \Delta t)$

  $\underline{\hat{u}}_m \leftarrow \underline{\hat{u}}_{m+1}$

  $\Delta\underline{\hat{u}} \leftarrow \underline{0}$

  $k \leftarrow 0$

  while ($k < k_{\max}$ **and** $||\underline{G}(\underline{\hat{u}}_m + \Delta\underline{\hat{u}})|| > \varepsilon$) **do** ▷ NEWTON ITERATION

    $\underline{\underline{K}} \leftarrow \underline{\underline{0}}$, $\underline{H} \leftarrow \underline{0}$, $\underline{q} \leftarrow \underline{0}$

    for $e = 1$: number of elements **do** ▷ assemble the system
      $\underline{\hat{u}}_e \leftarrow \underline{\underline{A}}_e \underline{\hat{u}}_m$ ▷ or a more efficient procedure
      ▷ instead of $\underline{\underline{A}}_e$

      $\Delta\underline{\hat{u}}_e \leftarrow \underline{\underline{A}}_e \Delta\underline{\hat{u}}$

      $\underline{\underline{K}}_e, \underline{H}_e, \underline{q}_e \leftarrow$ ElementRoutine($\underline{\hat{u}}_e$, $\Delta\underline{\hat{u}}_e$, ...)

      $\underline{\underline{K}} \leftarrow \underline{\underline{K}} + \underline{\underline{A}}^{\mathrm{T}}_e \underline{\underline{K}}_e \underline{\underline{A}}_e$

      $\underline{H} \leftarrow \underline{H} + \underline{\underline{A}}^{\mathrm{T}}_e \underline{H}_e$

      $\underline{q} \leftarrow \underline{q} + \underline{\underline{A}}^{\mathrm{T}}_e \underline{q}_e$

    end for

    $\underline{\underline{K}}_{\mathrm{c}}, \underline{H}_{\mathrm{c}}, \underline{q}_{\mathrm{c}}, \underline{\hat{F}}_{\mathrm{c}} \leftarrow$ EssBCs($\underline{\underline{K}}$, $\underline{H}$, $\underline{q}$, $\underline{\hat{F}}$, ...) ▷ essential boundary conditions

    $\blacktriangle\Delta\underline{\hat{u}} \leftarrow -\underline{\underline{K}}_{\mathrm{c}}^{-1}\left[\underline{H}_{\mathrm{c}} - \underline{q}_{\mathrm{c}} - \underline{\hat{F}}_{\mathrm{c}}\right]$

    $\Delta\underline{\hat{u}} \leftarrow \Delta\underline{\hat{u}} + \blacktriangle\Delta\underline{\hat{u}}$

    $k \leftarrow k + 1$

  end while
  if ($k \geq k_{\max}$ **and** $||\underline{G}(\underline{\hat{u}}_m + \Delta\underline{\hat{u}})|| > \varepsilon$) **then**
    conv $\leftarrow$ false ▷ no convergence
  else
    $\underline{\hat{u}}_{m+1} \leftarrow \underline{\hat{u}}_m + \Delta\underline{\hat{u}}$
    conv $\leftarrow$ true
  end if
  $t \leftarrow t + \Delta t$
end while

complexity of corresponding Finite-Element solution schemes. Some of the standard texts on non-linear FEM in Continuum Mechanics and Structural Analysis are [7–10]. Finite Element Methods in Fracture Mechanics are laid out, for instance, in [11].

References

1. S. Nakamura, *Applied Numerical Methods with Software*. Prentice-Hall international editions (Prentice Hall, 1991)
2. S. Barnard, J. Child, *Higher Algebra* (New Academic Science, 2017)
3. F.W.J. Olver, D.W. Lozier, R.F. Boisvert, C.W. Clark, *The NIST Handbook of Mathematical Functions* (Cambridge University, Press, 2010)
4. R. Ogden, *Non-linear Elastic Deformations*. Dover Civil and Mechanical Engineering (Dover Publications, 1997)
5. J. Lubliner, *Plasticity Theory*. Dover Books on Engineering (Dover Publications, 2013)
6. W. Brocks, *Plasticity and Fracture*. Solid Mechanics and Its Applications (Springer International Publishing, 2017)
7. W. Liu, B. Moran, T. Belytschko, K. Elkhodary, *Nonlinear Finite Elements for Continua and Structures* (Wiley, New York, 2013)
8. P. Wriggers, *Nonlinear Finite Element Methods* (Springer, Berlin Heidelberg, 2008)
9. J. Simo, T. Hughes, *Computational Inelasticity* (Interdisciplinary Applied Mathematics (Springer, New York, 2006)
10. J. Bonet, R. Wood, *Nonlinear Continuum Mechanics for Finite Element Analysis* (Cambridge University Press, 1997)
11. M. Kuna, *Finite Elements in Fracture Mechanics: Theory—Numerics—Applications*. Solid Mechanics and Its Applications (Springer, Netherlands, 2013)

Chapter 5
A Primer on Non-linear Dynamics and Multiphysics

Abstract This chapter combines the information presented in the two preceding chapters to present a solution procedure for non-linear dynamics. Phenomena or processes modelled by a system of coupled initial boundary value problems are usually classified as multi-physics problems. The development of FEM solution schemes for such cases is demonstrated for thermo-mechanical coupling. Since this chapter aims for a general understanding, only spatially one-dimensional problems are discussed.

5.1 Non-linear Dynamics

5.1.1 Strong and Time Continuous Variational Form

Considering only one spatial dimension, the strong form for non-linear structural dynamics reads

$$\rho\ddot{u} + \alpha\dot{u} - P' - r = 0 \quad x \in \Omega = (0, l) \tag{5.1}$$

with density ρ and damping function α. In the following, a non-linear relation between axial force P and strain $\epsilon = u'$ is assumed. Analogously to Sect. 3.5, (5.1) is second order in time and also second order in x. Therefore, two initial conditions and two boundary conditions are required. Exemplary, here the boundary conditions

$$u(t, 0) = 0 \tag{5.2}$$

$$P(t, l) = P_l(t) \tag{5.3}$$

with given function $P_l(t)$ and the initial conditions

$$u(0, x) = 0 \tag{5.4}$$

$$\dot{u}(0, x) = 0 \tag{5.5}$$

U. Mühlich, *Enhanced Introduction to Finite Elements for Engineers*, Solid Mechanics and Its Applications 268, https://doi.org/10.1007/978-3-031-30422-4_5

are used again for illustration purposes.

The time-continuous variational form is derived analogously to the previous sections. The details are omitted here. The result reads

$$\int_{\Omega} \rho \ddot{u}\, v\, \mathrm{d}x + \int_{\Omega} \alpha \dot{u}\, v\, \mathrm{d}x + \int_{\Omega} P v' \mathrm{d}x - \int_{\Omega} r\, v\, \mathrm{d}x - g(t) v(l) = 0 \tag{5.6}$$

$$u(t,0) = 0$$
$$\dot{u}(0,x) = 0$$
$$u(0,x) = 0$$

and the corresponding weak form is obtained by interpreting all spatial derivatives as weak derivatives

5.1.2 Galerkin FEM and Time Integration

Linear Lagrange elements are used exemplary and the FEM discretisation follows the same steps as in Sect. 3.4.2. Therefore, the contribution of a domain Ω_e, expressed in terms of quantities defined for $\Omega_\square$, is identical to (3.61), except for the product of element stiffness matrix and nodal displacements, which has to be replaced by $\underline{\boldsymbol{H}}_e$ given by (4.8), i.e.,

$$\mathcal{W}_{e\square} = \hat{\underline{v}}_\square^{\mathrm{T}} \left[\underline{\underline{\boldsymbol{M}}}_e \ddot{\hat{\underline{u}}}_\square + \underline{\underline{\boldsymbol{S}}}_e \dot{\hat{\underline{u}}}_\square + \underline{\boldsymbol{H}}_e - \underline{\boldsymbol{q}}_e \right] \tag{5.7}$$

with element mass matrix $\underline{\underline{\boldsymbol{M}}}_e$, element damping matrix $\underline{\underline{\boldsymbol{S}}}_e$, and $\underline{\boldsymbol{H}}_e$ given by (3.60), (3.46), and (4.8). Furthermore, the volume load vector $\underline{\boldsymbol{q}}_e$ is the same as in Sect. 2.1.7.

Taking into account the relations between local and global nodal values, for instance by using gathering matrices, leads eventually the following unconstrained system of equations

$$\underline{\underline{\boldsymbol{M}}}\, \ddot{\hat{\underline{u}}} + \underline{\underline{\boldsymbol{S}}}\, \dot{\hat{\underline{u}}} + \underline{\boldsymbol{H}}(\hat{\underline{u}}) = \underline{\boldsymbol{q}} + \hat{\underline{\boldsymbol{F}}} \tag{5.8}$$

from which the constrained system

$$\underline{\underline{\boldsymbol{M}}}\, \ddot{\hat{\underline{u}}} + \underline{\underline{\boldsymbol{S}}}\, \dot{\hat{\underline{u}}} + \underline{\boldsymbol{H}}(\hat{\underline{u}}) = \underline{\boldsymbol{q}} + \hat{\underline{\boldsymbol{F}}} - \underline{\boldsymbol{\Gamma}}^{\mathrm{D}} \tag{5.9}$$

is obtained by incorporating essential boundary conditions. Non-homogeneous Dirichlet boundary conditions are accounted for by $\underline{\boldsymbol{\Gamma}}^{\mathrm{D}}$. For the details see, for instance, Example 3.2.

Evaluating the constrained system at time $t = t_m$, approximating first time derivatives by means of central differences given in (3.19) and second derivatives using the difference quotient given by (3.20), yields

$$\underline{\underline{M}}\frac{1}{h^2}\left[\underline{\hat{u}}_{m+1} - 2\underline{\hat{u}}_m + \underline{\hat{u}}_{m-1}\right] + \underline{\underline{S}}\frac{1}{2h}\left[\underline{\hat{u}}_{m+1} + \underline{\hat{u}}_{m-1}\right] + \underline{H}(\underline{\hat{u}}_m) = \underline{q}_m + \underline{\hat{F}}_m - \underline{\Gamma}^{\mathrm{D}}_m$$

and, after rearranging

$$\underline{\underline{K}}\,\underline{\hat{u}}_{m+1} = -\underline{H}(\underline{\hat{u}}_m) + 2\underline{\underline{M}}\,\underline{\hat{u}}_m - \left[\underline{\underline{M}} - \frac{h}{2}\underline{\underline{S}}\right]\underline{\hat{u}}_{m-1} + \underline{q}_m h^2 + \underline{\hat{F}}_m h^2 - \underline{\Gamma}^{\mathrm{D}} h^2 \tag{5.10}$$

is obtained, with

$$\underline{\underline{K}} = \underline{\underline{M}} + \frac{h}{2}\underline{\underline{S}}\,.$$

It should be noted, that (5.10) is an explicit time integration scheme.

A particular advantage of the central difference method is that it ensures conservation of energy for non-linear problems. It is important to note, that not all time integration methods possess this property, see, e.g., [1] and references cited therein.

5.2 Thermo-Mechanical Coupling

5.2.1 Strong Form

The term multi-physics refers to models accounting for two or more coupled physical processes or phenomena. Standard examples are piezoelectric materials. i.e., electro-mechanical coupling, electrochemical processes, or thermo-mechanics. The latter is used here to illustrate main features of corresponding FEM solution schemes.

Small axial deformation combined with thermal strains is considered in a domain $\Omega = (0, l)$ during a time interval $(0, t_{\mathrm{E}})$, using the coordinates x and t for spatial location and time, respectively. Denoting the axial displacement by u, the total strain

$$\epsilon = u' \tag{5.11}$$

is the sum of mechanical strain ϵ_{M} and thermal strain ϵ_{T}, i.e.,

$$\epsilon = \epsilon_{\mathrm{M}} + \epsilon_{\mathrm{T}}\,. \tag{5.12}$$

The latter are given by

$$P = \lambda\,\epsilon_{\mathrm{M}}\,, \qquad \epsilon_{\mathrm{T}} = \alpha_{\mathrm{T}}\,[\theta - \theta_0]\,, \tag{5.13}$$

with axial force P, stiffness λ, thermal expansion coefficient α_{T}, and temperature θ. A reference temperature $\theta_0 = 0$ is used to keep equations as simple as possible.

Recall that dynamic equilibrium in one dimension reads

$$\rho\ddot{u} + \alpha\dot{u} - P' - n = 0\,, \tag{5.1 revisited}$$

with damping function α. Taking into account, (5.11), (5.12) and (5.13), (5.1) takes the form

$$\rho\,\ddot{u} + \alpha\,\dot{u} - \lambda u'' + \lambda\,\alpha_{\mathrm{T}}\theta' - \lambda'[u' - \alpha_{\mathrm{T}}\theta] - n = 0\,. \tag{5.14}$$

The mechanical part (5.14) is combined with heat transport in one dimension. Assuming a constant elastic stiffness, $\lambda = \lambda_0$, the system of differential equations for thermo-mechanical coupling reads

$$\rho\,\ddot{u} + \alpha\,\dot{u} - \lambda_0 u'' + \lambda_0\,\alpha_{\mathrm{T}}\theta' - n = 0 \tag{5.15}$$

$$\gamma\,\dot{\theta} - \beta\theta'' + \alpha_{\mathrm{d}}\dot{u} - r = 0 \tag{5.16}$$

where β is the heat conduction coefficient and distributed heat sources / sinks are accounted for by r. Density multiplied by specific heat capacity and cross section area is denoted by γ and heat production due to damping is accounted for by α_{d}.

Remark 5.1 (*Nonlinear coupling*) If the elastic stiffness depends strongly on temperature, the term $\alpha_{\mathrm{T}}\,\lambda'\theta$ cannot be neglected which implies a non-linear coupling because $\lambda' = \frac{\mathrm{d}\lambda}{\mathrm{d}\theta}\theta'$.

Suitable boundary and initial conditions complete the model given so far by (5.15) and (5.16). Exemplary, the boundary conditions

$$u(0,t) = 0 \qquad \lambda_0 u'(l,t) = P_l(t) \tag{5.17}$$

are considered for the mechanical part, whereas for the thermal part

$$\theta(0,t) = 0 \qquad \beta\,\theta'(l,t) = -\,g(t) \tag{5.18}$$

are used with given function $g(t)$ for the inward heat flux. Furthermore, the initial conditions

$$\theta(x,0) = 0 \tag{5.19}$$

and

$$u(x,0) = 0 \qquad \dot{u}(x,0) = 0 \tag{5.20}$$

are employed.

5.2.2 Time Continuous Variational Form and FEM Discretisation

The time-continuous variational form is derived analogously to the previous sections. However, because there are now two differential equations for two unknown functions, different test functions are employed, which are called v and ψ. The result reads

$$\mathcal{W}^u = \int_\Omega \rho\, \ddot{u}\, \mathrm{d}x + \int_\Omega \alpha\, \dot{u}\, \mathrm{d}x + \int_\Omega \lambda u' v'\, \mathrm{d}x + \int_\Omega \alpha_\mathrm{T}\, \theta' v\, \mathrm{d}x - \int_\Omega n v\, \mathrm{d}x - P_l v(l) = 0$$

$$\mathcal{W}^\theta = \int_\Omega \gamma\, \dot{\theta}\, \psi\, \mathrm{d}x + \int_\Omega \beta\, \theta' \psi'\, \mathrm{d}x + \int_\Omega r\, \psi\, \mathrm{d}x + g(t)\, \psi(l) = 0$$

if the boundary conditions (5.17) and (5.18) are taken into account. The two variational forms can be added which gives

$$\mathcal{W} = \mathcal{W}^u + \mathcal{W}^\theta \tag{5.21}$$

and the corresponding weak form is obtained by interpreting spatial derivatives in a weak sense.

Exemplary, linear interpolation is used for all trial and test functions. Analogously to Sects. 2.1.7 and 3.4.2, a mesh is defined together with a reference domain $\Omega_\square$ and corresponding mappings χ_e. Evaluating the integrals in (5.21) piecewise for Ω with N_e finite elements gives

$$\mathcal{W} = \sum_{e=1}^{N_e} \mathcal{W}_e + \underline{\hat{F}}$$

with

$$\begin{aligned}\mathcal{W}_e = & \int_\Omega \rho\, \ddot{u}\, \mathrm{d}x + \int_\Omega \alpha\, \dot{u}\, \mathrm{d}x + \int_{\Omega_e} \lambda u' v'\, \mathrm{d}x + \int_{\Omega_e} \alpha_\mathrm{T}\, \theta' v\, \mathrm{d}x - \int_{\Omega_e} n v\, \mathrm{d}x \\ & + \int_{\Omega_e} \gamma\, \dot{\theta}\, \psi\, \mathrm{d}x + \int_{\Omega_e} \beta\, \theta' \psi'\, \mathrm{d}x + \int_{\Omega_e} r\, \psi\, \mathrm{d}x + g(t)\, \psi(l)\end{aligned} \tag{5.22}$$

where $\underline{\hat{F}}$ accounts for the flux terms at the boundaries.

Applying the methodology laid out in Sects. 2.1.7 and 3.4.2, trial and test functions u and v for the reference domain $\Omega_\square$ are given as follows

$$u_\square(\xi) = \underline{N}^\mathrm{T}\,\underline{\hat{u}}_\square\,, \quad u'_\square(\xi) = \underline{B}^\mathrm{T}\,\underline{\hat{u}}_\square\,, \quad \dot{u}_\square(\xi) = \underline{N}^\mathrm{T}\,\underline{\dot{\hat{u}}}_\square\,, \quad \ddot{u}_\square(\xi) = \underline{N}^\mathrm{T}\,\underline{\ddot{\hat{u}}}_\square\,, \tag{5.23}$$

$$v_\square(\xi) = \underline{N}^\mathrm{T}\,\underline{\hat{v}}_\square\,, \quad v'_\square(\xi) = \underline{B}^\mathrm{T}\,\underline{\hat{v}}_\square \tag{5.24}$$

whereas for the thermal part

$$\theta_\square(\xi) = \underline{N}^\mathrm{T}\,\underline{\hat{\theta}}_\square\,, \qquad \theta'_\square(\xi) = \underline{B}^\mathrm{T}\,\underline{\hat{\theta}}_\square \qquad \dot{\theta}_\square(\xi) = \underline{N}^\mathrm{T}\,\underline{\dot{\hat{\theta}}}_\square \tag{5.25}$$

$$\psi_\square(\xi) = \underline{N}^\mathrm{T}\,\underline{\hat{v}}_\square\,, \qquad \psi'_\square(\xi) = \underline{B}^\mathrm{T}\,\underline{\hat{v}}_\square \tag{5.26}$$

are used. Hatted quantities refer to nodes of the reference domain.

With (5.23)–(5.26), the contribution of a domain Ω_e to the weak form can be expressed concisely as

$$\mathcal{W}_{e_\square} = \underline{\hat{V}}^\mathrm{T}_\square \left[\underline{\underline{M}}_e\,\underline{\ddot{Y}}_\square + \underline{\underline{S}}_e\,\underline{\dot{Y}}_\square + \underline{\underline{K}}_e\,\underline{\hat{Y}}_\square - \underline{q}_e\right] \tag{5.27}$$

with

$$\underline{\hat{Y}}^\mathrm{T}_\square = \left[\underline{\hat{u}}^\mathrm{T}_\square\;\underline{\hat{\theta}}^\mathrm{T}_\square\right], \qquad \underline{\dot{\hat{Y}}}^\mathrm{T}_\square = \left[\underline{\dot{\hat{u}}}^\mathrm{T}_\square\;\underline{\dot{\hat{\theta}}}^\mathrm{T}_\square\right], \qquad \underline{\ddot{\hat{Y}}}^\mathrm{T}_\square = \left[\underline{\ddot{\hat{u}}}^\mathrm{T}_\square\;\underline{\ddot{\hat{\theta}}}^\mathrm{T}_\square\right],$$

and

$$\underline{\hat{V}}^\mathrm{T}_\square = \left[\underline{\hat{v}}^\mathrm{T}_\square\;\underline{\hat{\psi}}^\mathrm{T}_\square\right].$$

Each of the element matrices $\underline{\underline{M}}_e$, $\underline{\underline{S}}_e$ and $\underline{\underline{K}}_e$ in (5.27) consists of four sub-matrices,

$$\underline{\underline{M}}_e = \begin{bmatrix} \underline{\underline{M}}^{uu}_e & \underline{\underline{0}} \\ \underline{\underline{0}} & \underline{\underline{0}} \end{bmatrix}, \qquad \underline{\underline{S}}_e = \begin{bmatrix} \underline{\underline{0}} & \underline{\underline{0}} \\ \underline{\underline{S}}^{\theta u}_e & \underline{\underline{S}}^{\theta\theta}_e \end{bmatrix}, \qquad \underline{\underline{K}}_e = \begin{bmatrix} \underline{\underline{K}}^{uu}_e & \underline{\underline{K}}^{u\theta}_e \\ \underline{\underline{K}}^{\theta u}_e & \underline{\underline{K}}^{\theta\theta}_e \end{bmatrix}, \tag{5.28}$$

and off-diagonal sub-matrices indicate coupling. The matrix $\underline{\underline{M}}^{uu}_e$ takes the form

$$\underline{\underline{M}}^{uu}_e = \int_{\xi=-1}^{1} \rho\,\underline{N}\,\underline{N}^\mathrm{T}\,j_e^{-1}\,\mathrm{d}\xi\,,$$

where j_e is the Jacobian of the mapping χ_e. The sub-matrices of $\underline{\underline{S}}_e$ are given by

$$\underline{\underline{S}}^{\theta u}_e = \int_{\xi=-1}^{1} \alpha_\mathrm{d}\,\underline{N}\,\underline{N}^\mathrm{T}\,j_e^{-1}\,\mathrm{d}\xi\,, \qquad \underline{\underline{S}}^{\theta\theta}_e = \int_{\xi=-1}^{1} \gamma\,\underline{N}\,\underline{N}^\mathrm{T}\,j_e^{-1}\,\mathrm{d}\xi\,,$$

whereas the sub-matrices of $\underline{\underline{K}}_e$ read

$$\underline{\underline{K}}_e^{uu} = \int_{\xi=-1}^{1} \lambda \, \underline{B} \, \underline{B}^{\mathrm{T}} \, j_e^{-1} \, \mathrm{d}\xi \qquad \underline{\underline{K}}_e^{u\theta} = \int_{\xi=-1}^{1} \alpha_{\mathrm{T}} \, \underline{N} \, \underline{B}^{\mathrm{T}} \, j_e^{-1} \, \mathrm{d}\xi$$

$$\underline{\underline{K}}_e^{\theta u} = \underline{\underline{0}} \qquad \underline{\underline{K}}_e^{\theta\theta} = \int_{\xi=-1}^{1} \beta \, \underline{B} \, \underline{B}^{\mathrm{T}} \, j_e^{-1} \, \mathrm{d}\xi \,,$$

and $\underline{\underline{K}}_e^{\theta u}$ vanishes only, because coupling between heat transfer and strain is not considered here. Similarly, $\underline{q}_e$ in (5.27) consists of two parts

$$\underline{q}_e = \begin{bmatrix} \underline{q}_e^u \\ \underline{q}_e^\theta \end{bmatrix} = \begin{bmatrix} \int_{\xi=-1}^{1} n \, \underline{N} \, j_e^{-1} \, \mathrm{d}\xi \\ \int_{\xi=-1}^{1} r \, \underline{N} \, j_e^{-1} \, \mathrm{d}\xi \end{bmatrix} .$$

Taking into account the relations between local and global degrees of freedom, for instance by using gathering matrices, see Sect. 2.1.7, the unconstrained global system of equations

$$\underline{\underline{S}} \, \dot{\underline{\hat{Y}}} + \underline{\underline{K}} \, \underline{\hat{Y}} - \underline{q} - \underline{\hat{F}} = \underline{0} \tag{5.29}$$

is obtained because of the arbitrariness of the test function values at the nodes.

The constrained system of equations possesses the same structure as (5.29) in the absence of non-homogeneous Dirichlet boundary condition. Provided that degrees of freedom are ordered accordingly, the constrained version of (5.29) takes the form

$$\begin{bmatrix} \underline{\underline{M}} & \underline{\underline{0}} \\ \underline{\underline{0}} & \underline{\underline{0}} \end{bmatrix} \begin{bmatrix} \ddot{\underline{\hat{u}}} \\ \ddot{\underline{\hat{\theta}}} \end{bmatrix} + \begin{bmatrix} \underline{\underline{0}} & \underline{\underline{0}} \\ \underline{\underline{S}}^{\theta u} & \underline{\underline{S}}^{\theta\theta} \end{bmatrix} \begin{bmatrix} \dot{\underline{\hat{u}}} \\ \dot{\underline{\hat{\theta}}} \end{bmatrix} + \begin{bmatrix} \underline{\underline{K}}^{uu} & \underline{\underline{K}}^{u\theta} \\ \underline{\underline{0}} & \underline{\underline{K}}^{\theta\theta} \end{bmatrix} \begin{bmatrix} \underline{\hat{u}} \\ \underline{\hat{\theta}} \end{bmatrix} - \begin{bmatrix} \underline{q}^u \\ \underline{q}^\theta \end{bmatrix} - \begin{bmatrix} \underline{\hat{F}}^u \\ \underline{\hat{F}}^\theta \end{bmatrix} = \begin{bmatrix} \underline{0} \\ \underline{0} \end{bmatrix}$$

and a time integration method has to be selected for solving it. For details about time integration, see Chap. 3.

It is worth noting, that under certain conditions the implementation effort can be reduced significantly. For instance, if heat production due to damping can be neglected, i.e., $\underline{\underline{S}}^{u\theta} \approx \underline{\underline{0}}$, the thermal part can be solved separately during a given time increment. The result enters the mechanical part as additional term on the right hand side.

5.3 Bibliographical Remarks

Regarding non-linear dynamics and thermomechanics, the reader is referred to [2, 3], respectively. Several multi-physics problems are discussed in [4].

References

1. P. Wriggers, *Nonlinear Finite Element Methods* (Springer, Berlin, 2008)
2. J. Bonet, A. Gil, R. Wood, *Nonlinear Solid Mechanics for Finite Element Analysis: Dynamics* (Cambridge University Press, 2021)
3. T. Hsu, *The Finite Element Method in Thermomechanics* (Springer, Netherlands, 2012)
4. A. Logg, K.A. Mardal, G.N. Wells, et al., *Automated Solution of Differential Equations by the Finite Element Method* (Springer, Berlin, 2012)

Appendix A
Elements of Linear Algebra

Abstract The chapter summarizes concepts from linear algebra required in the main part of this book. After introducing the real vector space, dimension and basis of a vector space are discussed briefly. The concepts of norm, distance and inner product are motivated. In the remaining part of this chapter, the vector space $\mathbb{R}^N$ together with an orthonormal basis is considered because every vector space of finite dimension N can be identified with $\mathbb{R}^N$ by means of bijective linear mappings called isomorphisms. Linear mappings are encoded by tensors. Here, only Cartesian tensors, i.e., tensors expressed in terms of orthonormal base vectors, are considered. The connection between second order tensors and matrices are discussed in detail. Eventually, specific matrix properties such as determinant, rank and condition number are provided.

A.1 Preliminaries

Abstracting from details allows for treating different problems the same way as long as they share several essential characteristics. The concept of a set, whose members can be objects of any kind, is an example of such an abstraction. Although the reader is supposed to be familiar with the principal ideas related to sets and mappings, a limited number of definitions required in the sequel is provided.

Definition A.1 (*Cartesian product*) The Cartesian product between two sets A and B, written as $A \times B$, is a set which contains all ordered pairs (a, b) with $a \in A$, $b \in B$.

The concept of a mapping or map generalizes the idea of a function and the corresponding definition is given below.

Definition A.2 (*Mapping*) Given two sets D and C. A mapping M from D to C, written as $M : D \rightarrow C$, relates every element of D to at least one element of C. D is called domain whereas C is referred to as co-domain.

Since mappings are ubiquitous in mathematics, it is natural to strive to categorize them independently of the specific properties of the involved sets. A first classification is achieved by means of the following characteristics.

U. Mühlich, *Enhanced Introduction to Finite Elements for Engineers*, Solid Mechanics and Its Applications 268, https://doi.org/10.1007/978-3-031-30422-4

Definition A.3 (*Injective mapping*) A mapping $D \to C$ is called injective or one-to-one if every element of D is mapped to exactly one element of C.

Definition A.4 (*Surjective mapping*) A mapping $D \to C$ is called surjective or onto if at least one element of D corresponds to every element of C.

Definition A.5 (*Bijective mapping*) A mapping is called bijective if it is injective and surjective.

Sets as such are of limited use. The interesting objects are sets equipped with some structure, for instance, a topological structure, an algebraic structure like addition, an order relation, etc. Of particular interest are mappings which are structure preserving. Such mappings are called morphisms. For example, isomorphisms, to be discussed in more detail later, play an important role in Linear Algebra.

A.2 Elementary Algebraic Structures

Vector spaces require an algebraic structure called field which can be defined via the concept of a group, more specifically a commutative or abelian group.

Definition A.6 (*Commutative group*) A set X together with a binary operation "$*$" that combines any two elements $x, y \in X$ to form an element written as $x * y = y * x \in X$, such that the following requirements

1. there exists en element e called identity with $e * x = x$
2. for every $x \in X$ there exists an inverse x^{-1} such that $x * x^{-1} = e$
3. $x * (y * z) = (x * y) * z$

are satisfied, is called a commutative or abelian group.

Definition A.7 (*Field*) Given two binary operations called addition and multiplication with identity elements denoted as "0" and "1", respectively. A set X is a field if it forms a commutative group under addition and if all elements different from "0" form a commutative group under multiplication.

The rational numbers $\mathbb{Q}$, the real numbers $\mathbb{R}$ as well as the complex numbers $\mathbb{C}$ are fields. Since there exist in addition an order relation, $\mathbb{Q}$ and $\mathbb{R}$ are even ordered fields.

A.3 The Real Vector Space

The vector space concept captures the essential algebraic aspects of a countless number of different problems. Vectors spaces are defined over a field according to Definition A.7. A real vector space is defined over the field $\mathbb{R}$ as follows.

Table A.1 Properties of the operations defined for a vector space $\mathcal{V}$ with $\boldsymbol{a}, \boldsymbol{b}, \boldsymbol{c} \in \mathcal{V}$ and $\alpha, \beta \in \mathbb{R}$

	Summation	Multiplication with real number
(i)	$\boldsymbol{a} \oplus \boldsymbol{b} = \boldsymbol{b} \oplus \boldsymbol{a}$	$\alpha \odot (\beta \odot \boldsymbol{a}) = (\alpha\, \beta) \odot \boldsymbol{a}$
(ii)	$(\boldsymbol{a} \oplus \boldsymbol{b}) \oplus \boldsymbol{c} = \boldsymbol{a} \oplus (\boldsymbol{b} \oplus \boldsymbol{c})$	$\alpha \odot (\boldsymbol{a} \oplus \boldsymbol{b}) = \alpha \odot \boldsymbol{a} \oplus \alpha \odot \boldsymbol{b}$
(iii)	There exists an element $\boldsymbol{0}$ such that $\boldsymbol{a} \oplus \boldsymbol{0} = \boldsymbol{a}$	$(\alpha + \beta) \odot \boldsymbol{a} = \alpha \odot \boldsymbol{a} \oplus \beta \odot \boldsymbol{a}$
(iv)	To every element $\boldsymbol{a}$ there exists exactly one element $-\boldsymbol{a}$ such that $\boldsymbol{a} \oplus (-\boldsymbol{a}) = \boldsymbol{0}$ holds.	There exists an element 1, such that $1 \odot \boldsymbol{a} = \boldsymbol{a}$ holds.

Definition A.8 (*Real vector space*) A real vector space $\mathcal{V}$ is a set whose elements are called vectors for which two binary operations are defined:

- summation $\oplus : \mathcal{V} \times \mathcal{V} \rightarrow \mathcal{V}$, and
- multiplication with a real number (scaling) $\odot : \mathbb{R} \times \mathcal{V} \rightarrow \mathcal{V}$

with properties given in Table A.1.

The use of $\oplus$ and $\odot$ indicates that these operations are not necessarily summation and scalar multiplication in the usual sense. On the contrary, they can even mean graphical operations. However, the ordinary "+" on the left hand side of property (iii) of the multiplication expresses the sum of two real numbers. Similarly, $(\alpha\beta)$ in the first property of the multiplication is the common product of two real numbers.

To simplify the notation, "$\oplus$" is often replaced by "+" and $\lambda \boldsymbol{a}$ is used instead of $\lambda \odot \boldsymbol{a}$. This is commonly known as operator overloading.

In the following, operator overloading is used except for cases for which it seems necessary to make the difference with ordinary summation and multiplication explicit.

Definition A.9 (*Linear independence*) N vectors $\boldsymbol{g}_i, i = 1, .., N, \boldsymbol{g}_i \in \mathcal{V}$ are linearly independent if

$$\alpha_1 \boldsymbol{g}_1 + \alpha_2 \boldsymbol{g}_2 + \cdots + \alpha_N \boldsymbol{g}_N = \sum_{i=1}^{N} \alpha_i \boldsymbol{g}_i = \boldsymbol{0}$$

can only be fulfilled if all real coefficients α_i vanish, i.e., $\alpha_i = 0$.

Definition A.9 uses index notation. A closer look reveals, that notation can be significantly simplified by employing the so-called summation convention. The corresponding rules are provided in the following definition.

Definition A.10 (*Summation convention*) Indexing obeys the following rules:

(i) an index appears, at most, twice within a term;
(ii) indices which appear only once per term must be the same in every term of an expression;
(iii) if an index appears twice, it indicates summation;
(iv) summation indices can be chosen freely as long as this does not conflict with (i)–(iii).

By employing summation convention, the right hand side of the equation in Definition A.9 can be written concisely as $\alpha_i \boldsymbol{g}_i$ without any ambiguity or loss of information. The range of summation is usually clear from the context.

Definition A.11 (*Maximal set of linearly independent vectors*) A set consisting of N vectors $\{\boldsymbol{g}_i\}$, $i = 1, .., N$, $\boldsymbol{g}_i \in \mathcal{V}$, $i \in \mathbb{N}$ is called a maximal set of linearly independent vectors if it cannot be extended by any other element of $\mathcal{V}$ without violating Definition A.9.

Definition A.12 (*Dimension of a vector space*) The dimension of a finite dimensional vector space is defined as the size of the sets according to Definition A.11.

Definition A.13 (*Basis of a vector space*) Every maximal set of linearly independent vectors according to Definition A.11 forms a basis of the considered vector space. The members of such a set are called base vectors.

A vector space has at least dimension one, and it should be noted, that the zero vector cannot belong to the set of base vectors, since $\alpha\mathbf{0} = \mathbf{0}$ for any $\alpha \in \mathbb{R}$.

A.4 Inner Product, Norm and Metric

So far, only algebraic aspects have been discussed. An algebraic structure is, however, not sufficient for performing analysis. Defining fundamental concepts of analysis, like neighbourhood, continuity, convergence, etc., requires at least a topology. A metric, i.e., the generalization of the notion of distance, induces a topology.

Definition A.14 (*Metric*) Given a set X. A mapping $d : X \times X \rightarrow \mathbb{R}$ with the properties:

(i) $d(x, y) = 0$ implies $x = y$
(ii) $d(x, y) = d(y, x)$
(iii) $d(x, z) \leq d(x, y) + d(y, z)$

for all $x, y, z \in X$ is called a metric.

A norm, which is a generalization of the notion of length (or size), induces a metric. Therefore, normed vector spaces play an important role in analysis.

Definition A.15 (*Norm*) A norm on a vector space $\mathcal{V}$, written as $||\boldsymbol{w}||$ for $\boldsymbol{w} \in \mathcal{V}$, is a mapping $\mathcal{V} \to \mathbb{R}$ with the properties:

(i) $||\alpha \boldsymbol{v}|| = \alpha ||\boldsymbol{v}||$
(ii) $||\boldsymbol{u} + \boldsymbol{v}|| \leq ||\boldsymbol{u}|| + ||\boldsymbol{v}||$
(iii) $||\boldsymbol{v}|| = 0$ implies $\boldsymbol{u} = \boldsymbol{0}$

for $\alpha \in \mathbb{R}$ and $\boldsymbol{u}, \boldsymbol{v} \in \mathcal{V}$.

Finally, the inner product generalizes the notion of angles. In addition, it induces a norm. Therefore, vector spaces with inner product are metric spaces.

Definition A.16 (*Inner product*) A mapping $\mathcal{V} \times \mathcal{V} \to \mathbb{R}$ with the following properties:

(i) $\boldsymbol{a} \cdot \boldsymbol{b} = \boldsymbol{b} \cdot \boldsymbol{a}$
(ii) $(\alpha \boldsymbol{a} + \beta \boldsymbol{b}) \cdot \boldsymbol{c} = \alpha \boldsymbol{a} \cdot \boldsymbol{c} + \beta \boldsymbol{b} \cdot \boldsymbol{c}$
(iii) $\boldsymbol{a} \neq \boldsymbol{0} \Rightarrow \boldsymbol{a} \cdot \boldsymbol{a} > 0$

for all $\boldsymbol{a}, \boldsymbol{b}, \boldsymbol{c} \in \mathcal{V}$ and $\alpha, \beta \in \mathbb{R}$, is called inner product. A vector space equipped with an inner product is called inner product space or pre-Hilbert space.

Furthermore, in the presence of an inner product, the so-called metric coefficients g_{ij} can be defined by

$$g_{ij} = \boldsymbol{g}_i \cdot \boldsymbol{g}_j \,. \tag{A.1}$$

A basis $\{\boldsymbol{g}_i\}$ with the property

$$\boldsymbol{g}_i \cdot \boldsymbol{g}_j = \delta_{ij}$$

is called orthonormal, where the Kronecker symbol δ_{ij} is defined as follows.

Definition A.17 (*Kronecker symbol*) The Kronecker symbol δ_{ij} with indexes $i, j = 1, \ldots, N$ is defined by

$$\delta_{ij} = \begin{cases} 1 & \text{if} \quad i = j \\ 0 & \phantom{\text{if}} \quad i \neq j \end{cases} .$$

A.5 The Vector Space $\mathbb{R}^N$

The definitions introduced so far are valid even if the dimension of the vector space is not finite but infinite. However, in the following only finite dimensional vector spaces are considered.

$\mathbb{R}^2$, for instance, is a short-hand notation for the Cartesian product $\mathbb{R} \times \mathbb{R}$. Therefore, a member p of $\mathbb{R}^2$ is just a pair of real numbers, written as

$$p = (x_1, x_2)$$

and the $\mathbb{R}^N$ as such is just the set of all possible n-tupels of real numbers called points in the following. Vectors can be defined as point differences by means of a mapping.

Definition A.18 (*Vectors in* $\mathbb{R}^N$) A vector in $\mathbb{R}^N$ is the result of a mapping F

$$F : \mathbb{R}^N \times \mathbb{R}^N \to \mathcal{V}$$
$$(p, q) \mapsto \boldsymbol{v}$$

where the ith component of $\boldsymbol{v}$, v_i is computed by $v_i = q_i - p_i$.

Definition A.19 (*Standard basis of* $\mathbb{R}^N$) Standard base vectors of $\mathbb{R}^N$ are denoted by $\boldsymbol{e}_i$ in the following. According to the convention introduced above, every $\boldsymbol{e}_i$ is a column with n elements. All elements are zero except the kth element which is equal to one.

The standard basis of $\mathbb{R}^N$ is an orthonormal basis. All following definitions and examples in this subsection refer to the standard basis.

Standard basis of $\mathbb{R}^2$

The standard basis of $\mathbb{R}^2$ is given by

$$\boldsymbol{e}_1 = \begin{bmatrix} 1 \\ 0 \end{bmatrix}, \qquad \boldsymbol{e}_2 = \begin{bmatrix} 0 \\ 1 \end{bmatrix},$$

and a vector $\boldsymbol{u}$ of $\mathbb{R}^2$ can be expressed in terms of the standard basis as

$$\boldsymbol{u} = u_1\boldsymbol{e}_1 + u_2\boldsymbol{e}_2 .$$

Following the relevant literature, $\mathbb{R}^N$ can refer either to the set of points or to the corresponding vector space. A likely reason for this abuse of notation might be, that the objects are in both cases n-tupels of real numbers.

However, the definition of vectors is not sufficient. In order for $\mathbb{R}^N$ to be a well defined vector space, it has to be endowed with two operations: summation of vectors and multiplication of a vector with a real number as defined in Definition A.8.

Definition A.20 (*Vector summation in* $\mathbb{R}^N$) The summation of two vectors $\boldsymbol{a}$ and $\boldsymbol{b}$ is written symbolically as $\boldsymbol{c} = \boldsymbol{a} + \boldsymbol{b}$. The ith component of $\boldsymbol{c}$ is given by $c_i = a_i + b_i$.

To make the difference between points and vectors somehow explicit by notation, vectors are written as columns, i.e., matrices with n rows and one column, in the following. In the case of $\mathbb{R}^2$, this means

$$a + b = \begin{bmatrix} a_1 \\ a_2 \end{bmatrix} + \begin{bmatrix} b_1 \\ b_2 \end{bmatrix} = \begin{bmatrix} a_1 + b_1 \\ a_2 + b_2 \end{bmatrix}.$$

Definition A.21 (*Multiplication of a vector with real number in* $\mathbb{R}^N$) The multiplication of a vector $\boldsymbol{a}$ with a real number β is written symbolically as $\boldsymbol{c} = \beta\boldsymbol{a}$. The i-th component of $\boldsymbol{c}$ is given by $c_i = \beta a_i$.

For instance, multiplication of a vector $\boldsymbol{a}$ with a real number β in $\mathbb{R}^2$ means

$$\beta\boldsymbol{a} = \begin{bmatrix} \beta a_1 \\ \beta a_2 \end{bmatrix}.$$

It is an easy task to verify, that the operations defined above obey the rules given in Table A.1. This is left as an exercise to the reader.

Instead of F, introduced in Definition A.18, a so-called hybrid addition can be used. It not only provides advantages in terms of notation but it allows as well for more transparent geometric interpretations.

Definition A.22 (*Hybrid addition of points and vectors in* $\mathbb{R}^N$) Hybrid addition "$\dotplus$" is a mapping

$$\begin{aligned} \dotplus : \mathbb{R}^N \times \mathbb{R}^N &\to \mathbb{R}^N \\ (p, \boldsymbol{u}) &\mapsto q = p \dotplus \boldsymbol{u} \end{aligned}$$

with the properties:

(i) $[p \dotplus \boldsymbol{u}] \dotplus \boldsymbol{v} = p \dotplus [\boldsymbol{u} + \boldsymbol{v}]$
(ii) $p \dotplus \boldsymbol{u} = p$ implies $\boldsymbol{u} = \boldsymbol{0}$
(iii) For all p, q, there exists a unique $\boldsymbol{u}$ such that $q = p \dotplus \boldsymbol{u}$.

Hybrid addition in $\mathbb{R}^2$

Given a point $\hat{p}$ in $\mathbb{R}^2$ with coordinates $(\hat{x}_1, \hat{x}_2)$, i.e., $p = (\hat{x}_1, \hat{x}_2)$. The coordinates of a point q given in terms of $\hat{p}$ and a vector $\boldsymbol{h}$ are computed by

$$(q_1, q_2) = \hat{p} \dotplus \boldsymbol{h} = (\hat{x}_1, \hat{x}_2) \dotplus \begin{bmatrix} h_1 \\ h_2 \end{bmatrix} = (\hat{x}_1 + h_1, \hat{x}_2 + h_2).$$

Therefore, hybrid addition allows for writing $f(\hat{x}_1 + h_1, \hat{x}_2 + h_2)$ simply as $f(\hat{p} \dotplus \boldsymbol{h})$.

Definition A.23 (*Inner product in* $\mathbb{R}^N$) The inner product of two vectors $\boldsymbol{a}$ and $\boldsymbol{b}$, written symbolically as $\boldsymbol{a} \cdot \boldsymbol{b}$ is given by

$$\boldsymbol{a} \cdot \boldsymbol{b} = a_i b_i \,.$$

The norm induced by this inner product is the so-called Euclidean norm

$$||\boldsymbol{a}|| = \sqrt{\boldsymbol{a} \cdot \boldsymbol{a}} = \sqrt{a_i a_i}$$

and the corresponding distance function between two vectors $\boldsymbol{a}$ and $\boldsymbol{b}$ reads

$$d(\boldsymbol{a}, \boldsymbol{b}) = ||\boldsymbol{a} - \boldsymbol{b}|| = \sqrt{a_i a_i - b_b b_i}$$

which is known as Euclidean distance.

A.6 Linear Mappings and Tensors

Definition A.24 (*Linear mapping between vector spaces*) Given two vector spaces $\mathcal{V}$ and $\mathcal{W}$. A mapping $\varphi : \mathcal{V} \to \mathcal{W}$ is linear if

$$\varphi(\alpha \boldsymbol{u} + \beta \boldsymbol{v}) = \alpha \varphi(\boldsymbol{u}) + \beta \varphi(\boldsymbol{v})$$

holds for any $\alpha, \beta \in \mathbb{R}$ and $\boldsymbol{u}, \boldsymbol{v} \in \mathcal{V}$.

To explore this concept further, the vector spaces $\mathcal{V}$ and $\hat{\mathcal{V}}$ are considered together with corresponding sets of orthonormal base vectors $\{\boldsymbol{g}_i\}$ and $\{\hat{\boldsymbol{g}}_i\}$, respectively. Furthermore, $\mathcal{V}$ and $\hat{\mathcal{V}}$ can have different dimensions. A mapping $A : \mathcal{V} \to \hat{\mathcal{V}}$ requires an element of $\mathcal{V}$, for instance $\boldsymbol{v}$, and it returns some vector $\boldsymbol{w}$ which belongs to $\hat{\mathcal{V}}$, i.e.,

$$\boldsymbol{w} = A(\boldsymbol{v}) = A(v_1 \boldsymbol{g}_1 + v_2 \boldsymbol{g}_2 + \cdots + v_n \boldsymbol{g}_n) = A(v_j \boldsymbol{g}_j)\,,$$

where $n = \dim(\mathcal{V})$. Linearity implies

$$\boldsymbol{w} = v_1 A(\boldsymbol{g}_1) + v_2 A(\boldsymbol{g}_2) + \cdots + v_n A(\boldsymbol{g}_n) = v_i A(\boldsymbol{g}_i) = v_i \boldsymbol{A}_i \,.$$

Since $\boldsymbol{w}$ is a vector of $\hat{\mathcal{V}}$ and the v_i are real numbers, the $\boldsymbol{A}_i$ must be vectors as indicated by notation. More specifically, the $\boldsymbol{A}_i$ are elements of $\hat{\mathcal{V}}$ and can therefore be written in terms of base vectors $\hat{\boldsymbol{g}}_i$ either as

$$\boldsymbol{A}_i = A_{ij} \hat{\boldsymbol{g}}_j \tag{A.2}$$

or as $\boldsymbol{A}_i = A_{ji} \hat{\boldsymbol{g}}_j$. Here, (A.2) is preferred which yields

$$\boldsymbol{w} = v_i A_{ij} \hat{\boldsymbol{g}}_i \,. \tag{A.3}$$

Splitting the mapping into an operator and an operand (argument) actually requires the dual space concept. However, in the presence of an inner product, the latter can be used instead. More specifically, for an orthonormal basis $\{\boldsymbol{g}_k\}$

$$v_i = \boldsymbol{v} \cdot \boldsymbol{g}_i = \boldsymbol{g}_i \cdot \boldsymbol{v}$$

holds and, therefore,

$$\boldsymbol{w} = (\boldsymbol{g}_i \cdot \boldsymbol{v})\, A_{ij}\, \hat{\boldsymbol{g}}_j = A_{ij}(\boldsymbol{g}_i \cdot \boldsymbol{v})\, \hat{\boldsymbol{g}}_j$$

because the order of the real numbers A_{ij} and $(\boldsymbol{g}_i \cdot \boldsymbol{v})$ is irrelevant for the result. Defining the object

$$\mathsf{A} = A_{ij}\boldsymbol{g}_i\hat{\boldsymbol{g}}_j\,, \tag{A.4}$$

equation (A.3) can be written symbolically as

$$\boldsymbol{w} = \mathsf{A} \cdot \boldsymbol{v}\,. \tag{A.5}$$

The object A is a so-called second order tensor. The name originates from the pairing of two base vectors. This pairing possesses the essential properties of a product, since it is associative and distributive with respect to vector addition. Therefore, it is also called tensor product or dyadic product. To emphasize this by notation, $\boldsymbol{g}_i \otimes \hat{\boldsymbol{g}}_j$ is commonly used instead of $\boldsymbol{g}_i\hat{\boldsymbol{g}}_j$. Here, this standard will be adopted as well. Furthermore, if the A_{ij} refer to an orthonormal basis, A is called a Cartesian tensor. A formal definition of the tensor product is given below.

Definition A.25 (*Tensor product*) The tensor product denoted by $\otimes$ is a mapping $\otimes : \mathcal{V} \times \hat{\mathcal{V}} \rightarrow \mathcal{V} \otimes \hat{\mathcal{V}}$ with properties

(i) $\boldsymbol{u} \otimes [\hat{\boldsymbol{v}} + \hat{\boldsymbol{w}}] = \boldsymbol{u} \otimes \hat{\boldsymbol{v}} + \boldsymbol{u} \otimes \hat{\boldsymbol{w}}$
(ii) $[\boldsymbol{u} + \boldsymbol{v}] \otimes \hat{\boldsymbol{w}} = \boldsymbol{u} \otimes \hat{\boldsymbol{w}} + \boldsymbol{v} \otimes \hat{\boldsymbol{w}}$
(iii) $\boldsymbol{u} \otimes [\lambda \hat{\boldsymbol{v}}] = \lambda \boldsymbol{u} \otimes \hat{\boldsymbol{v}}$
(iv) $\hat{\boldsymbol{u}} \otimes [\lambda \boldsymbol{v}] = \lambda \hat{\boldsymbol{u}} \otimes \boldsymbol{v}$

where $\boldsymbol{u}, \boldsymbol{v} \in \mathcal{V}$, $\hat{\boldsymbol{u}}, \hat{\boldsymbol{v}}, \hat{\boldsymbol{w}} \in \hat{\mathcal{V}}$ and $\lambda \in \mathbb{R}$.

It is important to note, that elements of $\mathcal{V} \otimes \hat{\mathcal{V}}$ are in general represented by linear combinations of tensor products.

> Important

Tensors can be used as operators to encode linear mappings. Different notations exist which usually involve as well specific conventions.

Tensor product spaces can be themselves vector spaces as the following example shows which can be seen as a particular manifestation of the ubiquity of the vector space concept.

The vector space $\mathbb{R}^N \otimes \mathbb{R}^m$

The set of all second order tensors, $\mathbb{R}^N \otimes \mathbb{R}^m$ with summation and scaling by a real number λ defined by

$$\begin{aligned} \mathsf{A} + \mathsf{B} &= A_{kl}\boldsymbol{e}_k \otimes \hat{\boldsymbol{e}}_l + B_{kl}\boldsymbol{e}_k \otimes \hat{\boldsymbol{e}}_l = [A_{kl} + B_{kl}]\boldsymbol{e}_k \otimes \hat{\boldsymbol{e}}_l \\ \lambda\mathsf{A} &= \lambda\, A_{kl}\boldsymbol{e}_k \otimes \hat{\boldsymbol{e}}_l \end{aligned}$$

is a vector space with dimension $n + m$. The standard basis of $\mathbb{R}^N \otimes \mathbb{R}^m$ is given by $\{\boldsymbol{e}_i \otimes \hat{\boldsymbol{e}}_k\}$, $i = 1, \ldots n$, $j = 1, \ldots, m$, where $\{\boldsymbol{e}_j\}$ and $\{\hat{\boldsymbol{e}}_j\}$ denote the sets of standard base vectors for $\mathbb{R}^N$ and $\mathbb{R}^m$, respectively. For instance, the elements of the standard basis of $\mathbb{R}^2 \otimes \mathbb{R}^2$ are $\boldsymbol{e}_1 \otimes \boldsymbol{e}_1$, $\boldsymbol{e}_1 \otimes \boldsymbol{e}_2$, $\boldsymbol{e}_2 \otimes \boldsymbol{e}_1$, and $\boldsymbol{e}_2 \otimes \boldsymbol{e}_2$, which shows $\dim(\mathbb{R}^2 \otimes \mathbb{R}^2) = 4$.

Furthermore, tensor product spaces can be endowed with an inner product as illustrated by the following example.

Inner product in $\mathbb{R}^N \otimes \mathbb{R}^m$

The inner product between two elements of $\mathbb{R}^N \otimes \mathbb{R}^m$, A and B is defined by

$$\mathsf{A} : \mathsf{B} = [A_{ij}\boldsymbol{e}_i \otimes \hat{\boldsymbol{e}}_j] : [B_{kl}\boldsymbol{e}_k \otimes \hat{\boldsymbol{e}}_l] = A_{ij}B_{kl}\delta_{ik}\delta_{jl} = A_{ij}B_{ij}\,.$$

The exercise to prove, that the inner product defined in Example A.6 possesses al necessary properties defined in Definition A.16 is left to the reader.

In order to change the order of factors of inner product operations, the transpose of a second order tensor is required.

Definition A.26 (*Transpose of a second order tensor*) The transpose A^{T} of a second order tensor A is defined by

$$\boldsymbol{u} \cdot [\mathsf{A} \cdot \boldsymbol{v}] = \boldsymbol{v} \cdot [\mathsf{A}^{\mathrm{T}} \cdot \boldsymbol{u}]\,.$$

Transpose of a second order tensor from $\mathbb{R} \otimes \mathbb{R}$

It follows from Definition A.26, that the transpose of a second order tensor $\mathsf{A} = A_{ij}\boldsymbol{e}_i \otimes \boldsymbol{e}_j$ is given by

$$\mathsf{A}^{\mathrm{T}} = [A_{ij}\boldsymbol{e}_i \otimes \boldsymbol{e}_j]^{\mathrm{T}} = A_{ji}\boldsymbol{e}_i \otimes \boldsymbol{e}_j = A_{ij}\boldsymbol{e}_j \otimes \boldsymbol{e}_i \,.$$

For the sake of completeness, the cross product is introduced next by means of the Levi-Civita symbol.

Definition A.27 (*Levi-Civita symbol*) The Levi-Civita symbol ϵ_{ijk} is defined as

$$\epsilon_{ijk} = \begin{cases} +1 & (i, j, k) \text{ is an even permutation of } (1, 2, 3) \\ -1 & (i, j, k) \text{ is an odd permutation of } (1, 2, 3) \\ 0 & \text{else} \end{cases}$$

with $i, j, k = 1, 2, 3$, where even permutations of (1, 2, 3) are (3, 1, 2), (2, 3, 1) and (1, 2, 3) itself. Odd permutations are (3, 2, 1), (1, 3, 2) and (2, 1, 3).

Definition A.28 (*Cross product*) The cross product of two vectors $\boldsymbol{u}, \boldsymbol{v} \in \mathcal{V}$ with $\dim(\mathcal{V}) = 3$ is defined in terms of an orthonormal basis $\{\boldsymbol{g}_k\}$ as

$$\boldsymbol{u} \times \boldsymbol{v} = u_i v_j \boldsymbol{g}_i \times \boldsymbol{g}_j = \epsilon_{ijk}\boldsymbol{g}_k \,.$$

The exterior product generalizes the idea of the cross product. Contrary to the latter, which only exists in three dimensions, the exterior product can be defined for any finite dimension.

Definition A.29 (*Exterior product*) The exterior product of two vectors $\boldsymbol{u}$ and $\boldsymbol{v}$ denoted by $\boldsymbol{u} \wedge \boldsymbol{v}$ is defined as

$$\boldsymbol{u} \wedge \boldsymbol{v} = \boldsymbol{u} \otimes \boldsymbol{v} - \boldsymbol{v} \otimes \boldsymbol{u} \,.$$

It has the following properties

$$\begin{array}{ll} \text{(i)} & \boldsymbol{u} \wedge \boldsymbol{v} = -\boldsymbol{v} \wedge \boldsymbol{u} \\ \text{(ii)} & [\boldsymbol{u} \wedge \boldsymbol{v}] \wedge \boldsymbol{w} = [\boldsymbol{u} \wedge \boldsymbol{v}] \wedge \boldsymbol{w} \\ \text{(iii)} & [\alpha\boldsymbol{u} + \beta\boldsymbol{v}] \wedge \boldsymbol{w} = \alpha\boldsymbol{u} \wedge \boldsymbol{w} + \beta\boldsymbol{v} \wedge \boldsymbol{w} \\ \text{(iv)} & \boldsymbol{u} \wedge [\alpha\boldsymbol{v} + \beta\boldsymbol{w}] = \alpha\boldsymbol{u} \wedge \boldsymbol{v} + \beta\boldsymbol{u} \wedge \boldsymbol{w} \end{array}$$

implied by the properties of the tensor product.

The exterior product between two vectors is also called a bivector. Definition A.29 indicates, that bivectors are actually second order tensors. More specifically, bivectors are skew-symmetric tensors. The set of all skew-symmetric tensors of a tensor product space $\mathbb{R}^N \otimes \mathbb{R}^N$ with $N \geq 2$ is a subspace of $\mathbb{R}^N \otimes \mathbb{R}^N$ called $\Lambda^2\mathbb{R}^N$. Elements of $\Lambda^2\mathbb{R}^N$ are of general form

$$\underset{\sim}{\mathsf{A}} = a_{ij}\boldsymbol{e}_i \wedge \boldsymbol{e}_j \,.$$

Furthermore, an inner product can be defined for $\Lambda^2\mathbb{R}^N$ as follows

$$\underset{\sim}{\mathsf{A}} : \underset{\sim}{\mathsf{B}} = a_{ij}\boldsymbol{e}_i \wedge \boldsymbol{e}_j : b_{kl}\boldsymbol{e}_k \wedge \boldsymbol{e}_l = a_{ij}b_{kl}\delta_{ik}\delta_{jl} = a_{ij}b_{ij}$$

which induces the norm

$$||\underset{\sim}{\mathsf{A}}|| = \sqrt{\underset{\sim}{\mathsf{A}} : \underset{\sim}{\mathsf{A}}}\,. \tag{A.6}$$

Bivectors are extremely useful to define rotations and oriented area elements. Given two vectors $\boldsymbol{a}$ and $\boldsymbol{b}$. If $\boldsymbol{a}, \boldsymbol{b} \in \mathrm{R}^2$, the corresponding area element is given by

$$\underset{\sim}{\mathsf{A}} = \boldsymbol{u} \wedge \boldsymbol{v} = u_1 v_2 \boldsymbol{e}_1 \wedge \boldsymbol{e}_2 + u_2 v_1 \boldsymbol{e}_2 \wedge \boldsymbol{e}_1$$

where the order of the exterior product determines its orientation. Taking into account, that $\boldsymbol{e}_1 \wedge \boldsymbol{e}_2 = -\boldsymbol{e}_2 \wedge \boldsymbol{e}_1$, see Definition A.29 (i),

$$\underset{\sim}{\mathsf{A}} = [u_1 v_2 - u_2 v_1]\boldsymbol{e}_1 \wedge \boldsymbol{e}_2$$

is obtained where the term with the brackets is the determinant of the matrix $[\boldsymbol{u}\ \ \boldsymbol{v}]$. The area μ can be computed by means of the norm (A.6), which gives

$$\mu\,(\boldsymbol{u} \wedge \boldsymbol{v}) = |u_1 v_2 - u_2 v_1|\,.$$

If, however, $\boldsymbol{a}, \boldsymbol{b} \in \mathrm{R}^3$, the corresponding area element in $\mathbb{R}^3$ is given by

$$\begin{aligned}\underset{\sim}{\mathsf{A}} = \boldsymbol{u} \wedge \boldsymbol{v} = \;& u_1 v_2 \boldsymbol{e}_1 \wedge \boldsymbol{e}_2 + u_2 v_3 \boldsymbol{e}_2 \wedge \boldsymbol{e}_3 + u_1 v_3 \boldsymbol{e}_1 \wedge \boldsymbol{e}_3 \\ & + u_2 v_1 \boldsymbol{e}_2 \wedge \boldsymbol{e}_1 + u_3 v_2 \boldsymbol{e}_3 \wedge \boldsymbol{e}_2 + u_3 v_1 \boldsymbol{e}_3 \wedge \boldsymbol{e}_1\,.\end{aligned}$$

Taking into account again (i) from Definition A.29 (i), it follows that

$$\underset{\sim}{\mathsf{A}} = \; [u_1 v_2 - u_2 v_1]\boldsymbol{e}_1 \wedge \boldsymbol{e}_2 + [u_2 v_3 - u_3 v_2]\boldsymbol{e}_2 \wedge \boldsymbol{e}_3 + [u_1 v_3 - u_3 v_1]\boldsymbol{e}_1 \wedge \boldsymbol{e}_3$$

and the corresponding area reads

$$\mu(\boldsymbol{u} \wedge \boldsymbol{v}) = \sqrt{[u_1 v_2 - u_2 v_1]^2 + [u_2 v_3 - u_3 v_2]^2 + [u_1 v_3 - u_3 v_1]^2}\,.$$

This scheme extends straight-forwardly to multiple wedge products and arbitrary dimensions. For instance, a volume element in $\mathbb{R}^3$ can be defined by

$$\boldsymbol{u} \wedge \boldsymbol{v} \wedge \boldsymbol{w} = \det\left([\boldsymbol{u}\ \boldsymbol{v}\ \boldsymbol{w}]\right) \boldsymbol{e}_1 \wedge \boldsymbol{e}_2 \wedge \boldsymbol{e}_3 \tag{A.7}$$

for $\boldsymbol{u}, \boldsymbol{v}, \boldsymbol{w} \in \mathbb{R}^3$. Its volume is given by $\mu(\boldsymbol{u} \wedge \boldsymbol{v} \wedge \boldsymbol{w}) = |\det\left([\boldsymbol{u}\ \boldsymbol{v}\ \boldsymbol{w}]\right)|$.

Splitting a mapping into a sequence of mappings with specific individual meanings like rotation or scaling is a rather common technique. The following example illustrates how to define proper rules for expressing sequences of linear mappings in terms of sequences of second order tensors.

Sequences of linear mappings in $\mathbb{R}^N$

Consider the operation

$$\boldsymbol{w} = A(B(\boldsymbol{u}))$$

where A and B are two different linear mappings from $\mathbb{R}^N$ to $\mathbb{R}^N$. Linearity implies

$$\boldsymbol{w} = A(B_{ij}u_j\boldsymbol{e}_i) = B_{ij}u_jA(\boldsymbol{e}_i) = B_{ij}u_jA(\boldsymbol{e}_i) = B_{ij}u_j\boldsymbol{A}_i = B_{ij}u_jA_{ki}\boldsymbol{e}_k$$

and taking into account, that $u_j = \boldsymbol{e}_j \cdot \boldsymbol{u}$ yields

$$\boldsymbol{w} = \left[A_{ki}B_{ij}\boldsymbol{e}_k \otimes \boldsymbol{e}_j\right] \cdot \boldsymbol{u} = \mathsf{C} \cdot \boldsymbol{u} \tag{A.8}$$

using the convention established in (A.5). It follows from (A.8), that

$$A(B(\boldsymbol{u})) = \mathsf{A} \cdot \mathsf{B} \cdot \boldsymbol{u} = \mathsf{C} \cdot \boldsymbol{u}\,.$$

The rule for the operation $\mathsf{C} = \mathsf{A} \cdot \mathsf{B}$ is given implicitly by (A.8). Furthermore,

$$[\mathsf{A}] \cdot [\mathsf{B} \cdot \boldsymbol{u}] = [\mathsf{A} \cdot \mathsf{B}] \cdot \boldsymbol{u}$$

holds. It is important to note, that the multiplication of tensors by means of the inner product does not commute, i.e., in general $\mathsf{A} \cdot \mathsf{B} \neq \mathsf{B} \cdot \mathsf{A}$.

The following example illustrates the versatility of the tensor concept together with the emergence of higher order tensors.

Higher order tensors

Consider a linear mapping between tensor product spaces, more specifically, $T : \mathbb{R}^N \otimes \mathbb{R}^N \rightarrow \mathbb{R}^N \otimes \mathbb{R}^N$, i.e.,

$$\mathsf{A} = T(\mathsf{B}) = B_{ij}T(\boldsymbol{e}_i \otimes \boldsymbol{e}_j)\,.$$

Since, the left hand side is a second order tensor, the $T(\boldsymbol{e}_i \otimes \boldsymbol{e}_j)$ must be second order tensors too. Hence

$$\mathsf{A} = B_{ij}\mathsf{T}_{ij} = B_{ij}T_{klij}\boldsymbol{e}_k \otimes \boldsymbol{e}_l = T_{klij}\boldsymbol{e}_k \otimes \boldsymbol{e}_l B_{ij}\,.$$

by setting $\mathsf{T}_{ij} = T(\boldsymbol{e}_i \otimes \boldsymbol{e}_j)$. The B_{ij} can be expressed as

$$B_{ij} = \boldsymbol{e}_i \cdot [\mathsf{B} \cdot \boldsymbol{e}_j] = \mathsf{B} : [\boldsymbol{e}_i \otimes \boldsymbol{e}_j] = [\boldsymbol{e}_i \otimes \boldsymbol{e}_j] : \mathsf{B} .$$

Therefore,

$$\mathsf{A} = [T_{klij}\boldsymbol{e}_k \otimes \boldsymbol{e}_l \otimes \boldsymbol{e}_i \otimes \boldsymbol{e}_j] : [B_{rs}\boldsymbol{e}_r \otimes \boldsymbol{e}_s] = \overset{<4>}{\mathsf{T}} : \mathsf{B}$$

with the forth order tensor $\overset{<4>}{\mathsf{T}}$. A typical example of a fourth-order tensor is the elasticity tensor which maps the strain tensor linearly to the stress tensor.

A.7 Linear Mappings and Matrices

Recall the linear mapping $A : \mathcal{V} \to \hat{\mathcal{V}}$ discussed at the beginning of the previous subsection together with its representation (A.3). Expressing the vector $\boldsymbol{w}$ in terms of an orthonormal basis of $\hat{\mathcal{V}}$, (A.3) can be written as

$$\left[w_i - \overline{A}_{ik} v_k\right] \hat{\boldsymbol{g}}_i = \boldsymbol{0} , \tag{A.9}$$

with $\overline{A}_{ik} = A_{ki}$. The base vectors $\{\hat{\boldsymbol{g}}\}$ are linearly independent. Therefore, Eq. (A.9) can only be fulfilled if

$$\overline{A}_{ik} v_k = w_i \tag{A.10}$$

holds with $i = 1, \ldots, n$ and $k = 1, \ldots, m$, since $\dim(\mathcal{V}) = n$ and $\dim(\hat{\mathcal{V}}) = m$. It is important to note, that up to this point no inner product is involved. The result (A.10) reveals, that a linear mapping $A : \mathcal{V} \to \hat{\mathcal{V}}$ is encoded by $n \times m$ real numbers. These numbers can be arranged naturally in a matrix with n rows and m columns.

Definition A.30 (*Matrix*) A $n \times m$ matrix $\underline{\underline{A}}$ consists of n rows and m columns. Its elements A_{ij} are indexed as follows. The first index specifies the row whereas the second index refers to the column. Matrices with only one column, i.e., $n \times 0$ matrices, are underlined only once, for instance, $\underline{\boldsymbol{w}}$.

Essential operations with matrices are summation, matrix multiplication and transposition. The corresponding definitions are given below.

Definition A.31 (*Matrix addition*) Given two $n \times m$ matrices $\underline{\underline{A}}$ and $\underline{\underline{B}}$. The elements of the sum $\underline{\underline{C}} = \underline{\underline{A}} + \underline{\underline{B}}$ are computed by $C_{ik} = A_{ik} + B_{ik}$.

Definition A.32 (*Matrix multiplication*) Given a $n \times m$ matrix $\underline{\underline{A}}$ and a $m \times p$ matrix $\underline{\underline{B}}$. The elements of the product $\underline{\underline{C}} = \underline{\underline{A}}\,\underline{\underline{B}}$ are computed by $C_{ij} = A_{ik} B_{kj}$.

Definition A.33 (*Matrix transposition*) Given a matrix $\underline{\underline{A}}$. Its transposed $\underline{\underline{A}}^{\mathrm{T}}$ is obtained by interchanging rows and columns of $\underline{\underline{A}}$.

Definition A.34 (*Symmetric matrix*) A matrix $\underline{\underline{A}}$ is symmetric if $\underline{\underline{A}} = \underline{\underline{A}}^{\mathrm{T}}$.

The outstanding importance of the vector space $\mathbb{R}^N$ is due to a concept called isomorphism which will be defined next.

Definition A.35 (*Isomorphism*) An isomorphism between vector spaces $\mathcal{V}$ and $\mathcal{W}$ is a bijective linear mapping. Vector spaces for which an isomorphism exists are called isomorphic, which is denoted by $\mathcal{V} \cong \mathcal{W}$.

All vector spaces with the same finite dimension are isomorphic. More specifically, all vector spaces with finite dimension N can be identified with $\mathbb{R}^N$. Informally speaking, it means that if you know one of them you know them all.

Linear Algebra in $\mathbb{R}^N$ is all about vector spaces $\mathbb{R}^N$ and linear mappings of type $\mathbb{R}^k \to \mathbb{R}^m$ encoded by $m \times k$ matrices.

The notation employed in the context of standard linear algebra is illustrated by the following example.

Linear mappings $\mathbb{R}^2 \to \mathbb{R}^2$

Using the standard basis of $\mathbb{R}^2$, a linear mapping $A : \mathbb{R}^2 \to \mathbb{R}^2$, which maps a vector $\boldsymbol{u}$ to a vector $\boldsymbol{w}$ can be written as

$$w_1 \begin{bmatrix} 1 \\ 0 \end{bmatrix} + w_2 \begin{bmatrix} 0 \\ 1 \end{bmatrix} = A_{1j} u_j \begin{bmatrix} 1 \\ 0 \end{bmatrix} + A_{2j} u_j \begin{bmatrix} 0 \\ 1 \end{bmatrix}$$

which gives

$$\begin{bmatrix} w_1 \\ w_2 \end{bmatrix} = \begin{bmatrix} A_{11}u_1 + A_{12}u_2 \\ A_{21}u_1 + A_{22}u_2 \end{bmatrix} = \begin{bmatrix} A_{11} & A_{12} \\ A_{21} & A_{22} \end{bmatrix} \begin{bmatrix} u_1 \\ u_2 \end{bmatrix} = \underline{\underline{A}}\,\underline{\boldsymbol{u}}\,. \tag{A.11}$$

Please note, that the right hand side of (A.11) is not an inner product but it encodes the pairing of a 2×2 matrix $\underline{\underline{A}}$ and a one column matrix $\underline{\boldsymbol{u}}$ according to Definition A.32.

The elements of the matrix $\underline{\underline{A}}$ in (A.11) are the components of a second order tensor A. However, not every matrix encodes a linear mapping. On the contrary, matrices are objects on their own right and matrix notation is often employed only for achieving compact notation. Therefore, a distinction by notation is recommendable, for instance between a one column matrix $\underline{v}$ and a vector $\boldsymbol{v}$.

A matrix representation like (A.11) is obviously not possible for higher-order tensors because the appearance of more than two indexes conflicts with the very definition of a matrix. Nevertheless, tensors of arbitrary order can always be mapped to second order tensors by means of proper isomorphisms as illustrated by the following example.

Isomorphism between $\mathbb{R}^2 \otimes \mathbb{R}^2$ and $\mathbb{R}^4$

To define a bijective mapping $\phi : \mathbb{R}^2 \otimes \mathbb{R}^2 \rightarrow \mathbb{R}^4$, an element A of $\mathbb{R}^2 \otimes \mathbb{R}^2$ is expressed in terms of the corresponding standard basis, see Example A.6,

$$\mathsf{A} = A_{11}\boldsymbol{e}_1 \otimes \boldsymbol{e}_1 + A_{12}\boldsymbol{e}_1 \otimes \boldsymbol{e}_2 + A_{21}\boldsymbol{e}_2 \otimes \boldsymbol{e}_1 + A_{22}\boldsymbol{e}_2 \otimes \boldsymbol{e}_2 \,.$$

A can be mapped to an element $\boldsymbol{a} \in \mathbb{R}^4$ given by

$$\boldsymbol{a} = A_{11}\hat{\boldsymbol{e}}_1 + A_{12}\hat{\boldsymbol{e}}_2 + A_{21}\hat{\boldsymbol{e}}_3 + A_{22}\hat{\boldsymbol{e}}_4 \,.$$

The mapping is linear, i.e., $\phi(\alpha\mathsf{A} + \beta\mathsf{B}) = \alpha\phi(\mathsf{A}) + \beta\phi(\mathsf{B})$, and, therefore, invertible. The example illustrates, that vector spaces of equal dimensions are isomorphic. Furthermore, there is not just one possible mapping to establish an isomorphism, but there are infinitely many.

Although, not all matrices are automatically related to linear mappings, many Linear Algebra textbooks tacitly refer to matrices as representations of linear mappings and even avoid mentioning explicitly the underlying concept of tensors. Since, components of second order tensors can be arranged in matrices, essential properties of second order tensors, such as rank, determinant, eigenvalues, orthogonality, etc., can be discussed by means of matrix properties.

A.8 Systems of Linear Equations and Matrix Properties

A system of linear equations can be expressed concisely in matrix notation as

$$\underline{\underline{A}}\,\underline{x} = \underline{b} \tag{A.12}$$

which is obviously a linear mapping of type (A.10). The objective is to determine the unknowns arranged in $\underline{x}$ for given $\underline{\underline{A}}$ and $\underline{b}$. A necessary but not sufficient condition for a unique solution to exist is that there are as many equations as unknowns, i.e., if $\underline{\underline{A}}$ is a square matrix. Here, only real square matrices are considered. The following questions need to be answered:

- Is there one unique solution?

- Are there bounds on the accuracy when computing solutions?

Since (A.12) encodes a linear mapping $\mathbb{R}^n \to \mathbb{R}^n$ where n equals the number of unknowns, $\underline{b}$ as well as the columns of $\underline{\underline{A}}$ can be interpreted as vectors of $\mathbb{R}^N$, i.e.,

$$\underline{\underline{A}} = \left[A_1 \; A_2 \; ... \; A_n \right] .$$

Therefore, (A.12) can be written as

$$x_1 A_1 + x_2 A_2 + \cdots + x_n A_n = \boldsymbol{b} . \tag{A.13}$$

which means, that $\boldsymbol{b}$ is expressed in terms of the A_i. However, to represent any given vector $\boldsymbol{b} \in \mathbb{R}^n$, the set $\{A_i\}$, $i = 1, .., n$, has to be a basis of the vector space $\mathbb{R}^n$, i.e., the A_i must be linearly independent. The number of linearly independent column vectors of a square matrix is called its rank.

Definition A.36 (*Rank of a square matrix*) The column rank of a square matrix $\underline{\underline{A}}$, indicated by $\text{rk}(\underline{\underline{A}})$, is given by the number of linearly independent column vectors. It equals the row rank, i.e., the number of linearly independent rows interpreted as vectors. A $m \times m$ matrix $\underline{\underline{A}}$ has full rank if $\text{rk}(\underline{\underline{A}}) = m$. Otherwise, it is rank deficient.

There are different ways to determine the rank of a matrix. On the other hand, if the A_i are not linearly independent,

$$\overset{<n>}{\mathsf{A}} = A_1 \wedge A_2 \wedge \cdots \wedge A_n = 0 \in \underbrace{\mathbb{R}^N \otimes \cdots \otimes \mathbb{R}^N}_{n-\text{times}} .$$

As illustrated by the examples below Definition A.29, there exists the following relation between exterior product and determinants

$$\overset{<n>}{\mathsf{A}} = \det(\underline{\underline{A}}) \boldsymbol{e}_1 \wedge \boldsymbol{e}_2 \wedge \cdots \wedge \boldsymbol{e}_n ,$$

from which it can be concluded, that $\{A_i\}$ with $i = 1, .., n$, can only be a basis of $\mathbb{R}^N$, if $\det(\underline{\underline{A}}) \neq 0$.

Furthermore, $\underline{x}$ in (A.12) can be determined by matrix inversion. In this context, the following definitions are required.

Definition A.37 (*Identity matrix*) The diagonal elements of an identity matrix are equal to one and all remaining elements are zero.

Definition A.38 (*Matrix inverse*) The inverse $\underline{\underline{A}}^{-1}$ of a square matrix $\underline{\underline{A}}$ is defined by

$$\underline{\underline{A}}^{-1} \underline{\underline{A}} = \underline{\underline{I}} ,$$

with identity matrix $\underline{\underline{I}}$. A matrix whose inverse exists is called non-singular or invertible.

$\underline{\underline{A}}^{-1}$ can be computed using its cofactor matrix $\underline{\underline{C}}_{\underline{\underline{A}}}$ as follows

$$\underline{\underline{A}}^{-1} = \frac{1}{\det(\underline{\underline{A}})} \underline{\underline{C}}_{\underline{\underline{A}}}^{\mathrm{T}} .$$

A precise definition of the cofactor matrix is not even necessary at this point. It suffices to know, that its elements are determinants of corresponding sub-matrices of $\underline{\underline{A}}$. Since the inverse of a square matrix is proportional to the reciprocal of its determinant, (A.12) can only be solved if $\det(\underline{\underline{A}}) \neq 0$.

Based on the results discussed so far, the homogeneous case of (A.12) can be examined easily. Inspecting (A.13) for $\boldsymbol{b} = \mathbf{0}$ reveals, that there is of course the trivial solution $x_i = 0$. Additional solutions can only exist if the $\boldsymbol{A}_i$ are linearly dependent, i.e., if $\det(\underline{\underline{A}}) = 0$. In this case, an infinite number of solutions exist because if $\underline{\hat{x}}$ solves $\underline{\underline{A}}\,\underline{x} = \underline{0}$, any multiple of $\underline{\hat{x}}$ is a solution too.

A vector, which is only affected by a matrix $\underline{\underline{A}}$ in terms of scaling, is called an eigenvector of $\underline{\underline{A}}$. This concept can be cast into the equation

$$\underline{\underline{A}}\,\underline{\eta} = \lambda \underline{\eta} \tag{A.14}$$

where λ is a real number. Making use of the properties of the identity matrix $\underline{\underline{I}}$, Eq. (A.14) can be written as

$$\left[\underline{\underline{A}} - \lambda \underline{\underline{I}}\right] \underline{\eta} = \underline{0}$$

to see, that non-trivial solutions can only exist, if

$$\det\left(\underline{\underline{A}} - \lambda \underline{\underline{I}}\right) = 0 \tag{A.15}$$

holds. Equation (A.15) defines the characteristic polynomial of degree m for a $m \times m$ matrix $\underline{\underline{A}}$. The roots of this polynomial are the eigenvalues λ_i of $\underline{\underline{A}}$. Some important facts about eigenvalues and specific matrices are:

- A square matrix whose eigenvalues are all positive is called positive-definite.
- Real symmetric matrices have only real eigenvalues.

Once the eigenvalues λ_i are known, the corresponding eigenvectors $\underline{\eta}_i$ can be determined. Eigenvectors of distinct eigenvalues are orthogonal. It is convenient to define normalized eigenvectors

$$\overline{\underline{\eta}}_i = \frac{1}{||\underline{\eta}_i||} \underline{\eta}_i \qquad \text{(no sum)}$$

where here $||.||$ indicates the Euclidean norm in $\mathbb{R}^N$. The normalized eigenvectors form a orthonormal basis of the so-called eigenspace, which reads in matrix notation

$$\underline{\overline{\eta}}_i^{\mathrm{T}} \underline{\overline{\eta}}_j = \delta_{ij} .$$

One particularly useful application of eigenvectors is the so-called eigen-decomposition of a $n \times n$ matrix $\underline{\underline{A}}$. Such a decomposition is only possible for matrices with n distinct eigenvalues. Defining a matrix $\underline{\underline{Q}}$ in terms of the nomalized eigenvectors of $\underline{\underline{A}}$ as follows

$$\underline{\underline{Q}} = \left[\underline{\overline{\eta}}_1 \; \underline{\overline{\eta}}_2 \cdots \underline{\overline{\eta}}_n\right]$$

it can be shown, that

$$\underline{\underline{A}}\,\underline{\underline{Q}} = \left[\lambda_1\underline{\overline{\eta}}_1 \; \lambda_2\underline{\overline{\eta}}_2 \cdots \lambda_n\underline{\overline{\eta}}_n\right]$$

and, therefore,

$$\underline{\underline{Q}}^{\mathrm{T}}\,\underline{\underline{A}}\,\underline{\underline{Q}} = \underline{\underline{\Lambda}}$$

with

$$\Lambda_{ij} = \begin{cases} \lambda_i & i = j \\ 0 & i \neq j \end{cases} .$$

Since, $\underline{\underline{\Lambda}}$ is a diagonal matrix, its inverse can be determined easily. The advantage in the context of systems of linear equations should be obvious.

Eventually, the numerical values for the unknowns of a system of linear equations are of interest. In this context, the question arises if there are bounds on the accuracy of the solution. This leads directly to the concept of condition number of a given matrix. Its meaning will be illustrated by considering the system

$$\underline{\underline{A}}\,\underline{x} = \underline{b} + \Delta\underline{b}$$

where $\Delta\underline{b}$ represents a deviation of the correct right-hand side vector $\underline{b}$. To estimate, how such a deviation affects the solution $\underline{x}$, the latter is expressed formally as

$$\underline{x} = \underline{\underline{A}}^{-1}\underline{b} + \underline{\underline{A}}^{-1}\Delta\underline{b}$$

in order to define the relative error in the solution by

$$\Delta_{\mathrm{rel}}\underline{x} = \frac{||\underline{\underline{A}}^{-1}\Delta\underline{b}||}{||\underline{\underline{A}}^{-1}\underline{b}||} .$$

The relative error in the right-hand side vector is defined as

$$\Delta_{\text{rel}}\underline{\boldsymbol{b}} = \frac{||\Delta\underline{\boldsymbol{b}}||}{||\underline{\boldsymbol{b}}||}$$

and the condition number κ is the ratio of the relative error in the solution to the relative error in $\underline{\boldsymbol{b}}$, i.e.,

$$\kappa = \frac{\Delta_{\text{rel}}\underline{\boldsymbol{x}}}{\Delta_{\text{rel}}\underline{\boldsymbol{b}}} = \frac{||\underline{\underline{\boldsymbol{A}}}^{-1}\Delta\underline{\boldsymbol{b}}||}{||\underline{\underline{\boldsymbol{A}}}^{-1}\underline{\boldsymbol{b}}||}\frac{||\underline{\boldsymbol{b}}||}{||\Delta\underline{\boldsymbol{b}}||}\,. \tag{A.16}$$

All norms used so far refer to a vector norm in $\mathbb{R}^N$, for instance, the Euclidean norm. Deriving a useful estimate for κ based on the properties of $\underline{\underline{\boldsymbol{A}}}$ is not possible without discussing matrix norms first, at least to a certain extent.

Real square matrices of same dimension, together with a summation and a multiplication by a real number form a real vector space. The vector space of all $m \times m$ matrices has dimension m^2, which can be seen easily by defining a basis for the vector space. A variety of norms can be defined which comply with Definition A.15. Of particular importance are matrix norms which are compatible with norms defined for the corresponding vector spaces. Compatibility means

$$||\underline{\underline{\boldsymbol{A}}}\,\underline{\boldsymbol{z}}|| \leq ||\underline{\underline{\boldsymbol{A}}}||_{\boxdot}\,||\underline{\boldsymbol{z}}|| \tag{A.17}$$

where $||.||_{\boxdot}$ indicates a compatible matrix norm. In this case, the maximum of the condition number can be estimated as

$$\max(\kappa) = ||\underline{\underline{\boldsymbol{A}}}^{-1}||_{\boxdot}||\underline{\underline{\boldsymbol{A}}}||_{\boxdot}$$

by using (A.16) and (A.17) which is left as an illustrative exercise to the reader.

A.9 Bibliographical Remarks

Besides the large amount of standard textbooks in Linear Algebra, [1] can be recommended as a modern compendium. In addition, the large number of historical remarks regarding important contributions made by different mathematicians, physisicst and engineers will most likely be welcomed by the reader. Excellent didactic introductions into the subject are provided, for instance, by [2] and [3]. A highly recommendable introduction regarding the exterior product can be found in [4].

The notation used here for Cartesian tensors follows in large parts [5]. In addition, [6, 7] are suitable sources regarding the subject, especially for engineers.

References

1. J. Golan, *The Linear Algebra a Beginning Graduate Student Ought to Know*. Dover books on mathematics (Springer, Berlin, 2007)
2. G. Strang, *Linear Algebra and Its Applications*, 4th edn. (Brooks/Cole, Florence, KY, 2005)
3. S. Axler, *Linear Algebra Done Right*, 3rd edn. Undergraduate texts in mathematics (Springer International Publishing, Cham, Switzerland, 2014)
4. S. Winitzki, *Linear Algebra Via Exterior Products* (lulu.com, 2010)
5. G. Holzapfel, *Nonlinear Solid Mechanics: A Continuum Approach for Engineering* (Wiley, New York, 2000)
6. M. Itskov, *Tensor Algebra and Tensor Analysis for Engineers: With Applications to Continuum Mechanics* (Springer, Berlin, 2009)
7. U. Mühlich, *Fundamentals of Tensor Calculus for Engineers with a Primer on Smooth Manifolds*. Solid Mechanics and Its Applications (Springer International Publishing, 2018)

Appendix B
Elements of Real Analysis

Abstract The chapter summarizes concepts from real analysis required in the main part of this book. Since real analysis deals with infinite processes, a short overview about convergence of infinite sequences and series is provided first. This overview includes the construction of the real numbers as completion of the rational numbers, continuity of real functions as well as convergence of infinite sequences and series of real functions to a limit. Differentiability in $\mathbb{R}$ is examined to prepare the passage from $\mathbb{R}$ to $\mathbb{R}^2$ which is in general the most challenging step. Concepts such as Gateaux differential, partial derivatives, differential operators, etc,. are discussed in detail mostly for $\mathbb{R}^2$ because the generalization of these concepts to $\mathbb{R}^N$ is straightforward. The Riemann integral is provided and the main ideas behind Lebesgue integration are outlined. Eventually, Gauss integration theorems are summarized and the effect of coordinate transformations on integration is outlined.

B.1 Basic Topological Aspects

Analysis requires some space, i.e., a set of points together with some structure which allows to define a topology.

Being connected, boundary, interior, exterior, continuity and compactness are some examples of topological concepts. Topological properties are those which do not change under so-called homeomorphisms, i.e., continuous mappings with continuous inverse as illustrated in Fig. B.1. Under a continuous mapping ϕ, metric properties such as curvature or length of a line may change whereas the points of the line stay connected. In addition, the square remains inside the circle although the shapes of these objects change.

A topology can be established by defining, for instance, which sets are considered as open. Such an abstract approach allows to focus on topology without being distracted by some additional structure. On the other hand, a topology can be induced by a metric. Hence, all metric spaces are automatically topological spaces. Usually,

U. Mühlich, *Enhanced Introduction to Finite Elements for Engineers*, Solid Mechanics and Its Applications 268, https://doi.org/10.1007/978-3-031-30422-4

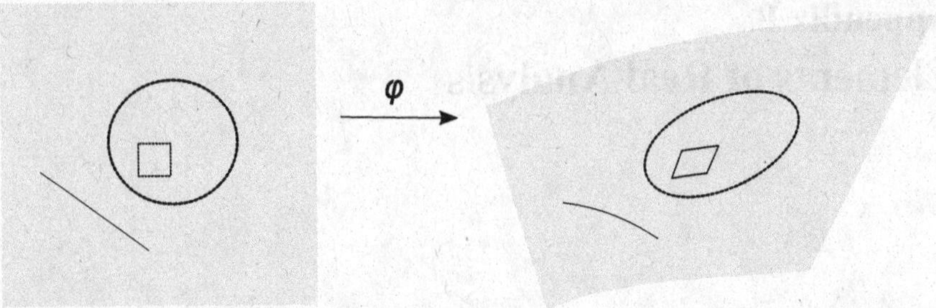

Fig. B.1 Illustration of a point set X together with the result of a continuous mapping. Solid lines indicate subsets of X defining geometric figures

topology in metric spaces is discussed in terms of an open ball the radius of which has length ε. However, length is not a topological property. Therefore, when working with metric spaces, it is not always immediately obvious if a concept is topological in nature without testing its behaviour under homeomorphisms.

B.2 Limits and Convergence of Sequences and Series

Fundamental concepts of analysis like derivative or integral rely on the notion of infinite processes encoded as infinite sequences and series. Since real computation is only possible for a finite number of operations, questions regarding convergence and limits of such processes play an essential role.

Definition B.1 (*Sequence of rational numbers*) A sequence of rational numbers, written as $\{r_n\}$, is a mapping $\mathbb{N} \to \mathbb{Q}$, more specifically $k \mapsto r_k$ with $k \in \mathbb{N}$, $r_k \in \mathbb{Q}$, generating an ordered list of rational numbers according to the order structure of $\mathbb{N}$.

Typical examples for sequences are $\{2^n\}$ and $\left\{\frac{1}{1+n}\right\}$. If necessary, sub-indexes and super-indexes are used to specify the domain, for example, $\{2^n\}_{n=1}^{4}$, $\left\{\frac{1}{1+n}\right\}_{n=1}^{\infty}$. The examples show, that sequences can be finite or infinite. The most useful sequences are those which converge to some limit value.

Definition B.2 (*Convergence of a rational sequence to a limit*) A sequence of rational numbers $\{r_n\}$ converges to a limit L, if for every $\epsilon > 0$ there exists an $N \in \mathbb{N}$ such that $|a_n - L| < \epsilon$ holds for every $n > N$.

Definition B.2 relies on a metric, more specifically the absolute value of the difference of two rational numbers. There exists as well a topological version which means, that convergence of rational sequences is actually a topological concept. The usefulness of Definition B.2 is rather restricted since the limit of a sequence has to be known or assumed beforehand. A more general way to check if a sequence converges is achieved by the following definition.

Definition B.3 (*Cauchy sequence of rational numbers*) A sequence $\{r_n\}$ is a Cauchy sequence if, for every $\epsilon > 0$, there exists a natural number N such that $|r_n - r_m| < \epsilon$ for any $n, m \geq N$, with $\epsilon \in \mathbb{Q}$ and $n, m \in \mathbb{N}$.

Cauchy sequences are bounded by definition. Furthermore, convergent sequences are Cauchy sequences. In addition, if a Cauchy sequence converges to a limit, this limit is unique.

A rather crucial observation is, that not all rational Cauchy sequences converge to rational numbers which is illustrated by the following example.

$\sqrt{2}$ as the results of a limit process

Rational numbers are solutions to certain algebraic equations. For instance, the solution of $x^2 - 4 = 0$ is $x = \sqrt{4} = 2$ and the square root is well defined for this case. However, $x^2 - 2 = 0$ has no rational solution. Writing $x = \sqrt{2}$ is meaningless in the context of $\mathbb{Q}$ since $\sqrt{2}$ can not be expressed as a fraction of two integer numbers, i.e., it is not a rational number.
On the other hand, one could intend to solve $x^2 - 2 = 0$ by means of an iterative process. Setting $x_0 = 1$ and defining $x_{n+1} = \frac{1}{2}[x_n + \frac{2}{x_n}]$ generates an infinite recursive sequence $\{x_n\} = \{1, 3/2, 17/12, \ldots\}$. It is a Cauchy sequence of rational numbers which has no limit in $\mathbb{Q}$. The iteration is known as Babylonian square-root algorithm.

This observation eventually leads to the concept of completeness together with the idea to construct the real numbers $\mathbb{R}$ as completion of $\mathbb{Q}$ by including all possible irrational limits of rational Cauchy sequences. This, however, requires some preparatory work.

Different Cauchy sequences can converge to the same limit, which makes it necessary to work with equivalence classes. Recall, that the construction of $\mathbb{Q}$ relies as well on the concept of equivalence classes. For instance, $1/2$ and $2/4$ have the same effect if used in operations like summation and multiplication. Therefore, they belong to the same equivalence class. Any member of this class can be used in computations to represent the rational number $1/2$.

Before defining equivalence classes of rational sequences the concept of zero sequences is required.

Definition B.4 (*Zero sequence of rational numbers*) A rational sequence $\{x_n\}$ is a zero sequence if for all $\epsilon > 0$ there is a natural number N such that $|x_n| < \epsilon$ for all $n > N$ with $n \in \mathbb{N}$ and $\epsilon \in \mathbb{Q}$.

Definition B.5 (*Equivalence of rational Cauchy sequences*) Two Cauchy sequences $\{x_n\}$, $\{y_n\}$ are equivalent if their difference is a zero sequence. Equivalence is expressed by $\{x_n\} \sim \{y_n\}$.

Given two sequences $\{x_n\}$, $\{y_n\}$ with

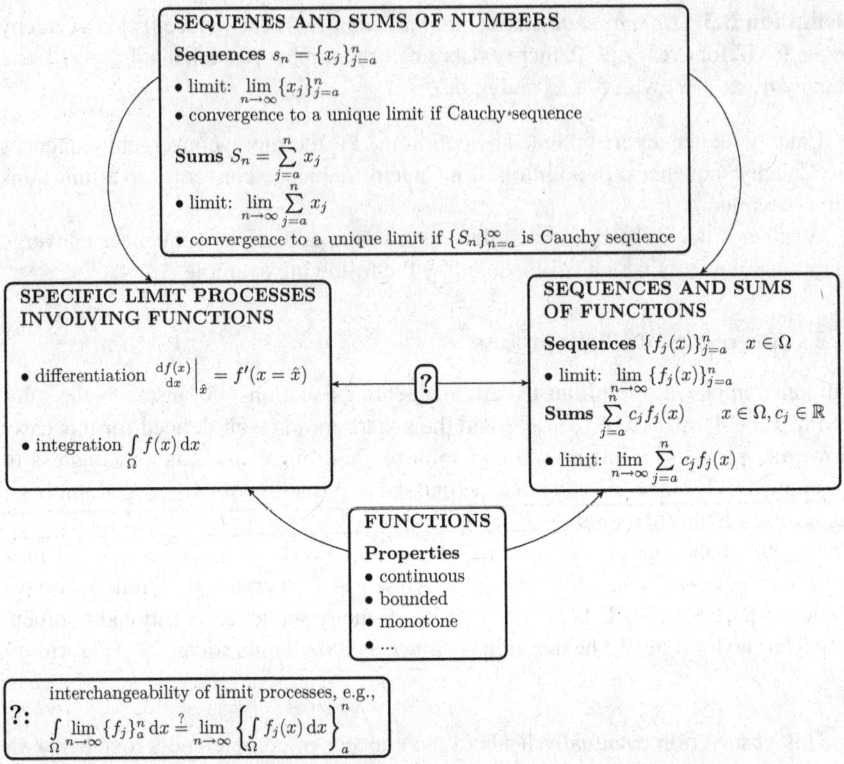

Fig. B.2 Schematic overview illustrating the importance of infinte sequences and infinite sums (series) in the context of real analysis

$$\lim_{n\to 0}\{x_n\} = a\,, \qquad \lim_{n\to 0}\{y_n\} = b$$

and $a, b \in \mathbb{Q}$, summation and multiplication of rational Cauchy sequences are defined as follows

$$\{x_n\} + \{y_n\} = a + b$$
$$\{x_n\}\{y_n\} = a\,b$$

which makes the set of equivalence classes of Cauchy sequences with limits in $\mathbb{Q}$ a field according to Definition A.7. The real numbers can be defined now as follows.

Definition B.6 (*Real numbers* $\mathbb{R}$) The set of all equivalence classes of rational Cauchy sequences is called the set of real numbers $\mathbb{R}$.

Definition B.6 implies, that $\mathbb{R}$ includes as well Cauchy sequences which do not converge to a rational number. The objects to which these Cauchy sequences converge are called irrational numbers. Irrational numbers are represented by corresponding infinite rational sequences, more specifically, by equivalence classes of corresponding sequences. $\mathbb{Q}$ is dense in $\mathbb{R}$ by construction according to the following definition.

Definition B.7 (*Dense subset*) A subset Y of X is dense in X if every point of X either belongs to Y or is limit point of Y.

The concept of Cauchy sequences can be extended to $\mathbb{R}$. Since every Cauchy sequence of real numbers has a limit in $\mathbb{R}$, $\mathbb{R}$ is a complete field. Furthermore, $\mathbb{R}$ inherits the order structure from $\mathbb{Q}$, i.e., $\mathbb{R}$ is a complete ordered field.

The concept of completeness defined in terms of Cauchy sequences can be generalized to any metric space.

Working with sequences requires some classification according to certain characteristics such as bounded and monotone. These characteristics are defined in the following for sequences of real numbers although some of them apply already for rational sequences.

Definition B.8 (*Bounded sequence*) A sequence $\{x_n\}$ is bounded, if there are numbers b_l, b_u for which $b_l \leq x_n \leq b_u$ holds for any n with $n \in \mathbb{N}$ and $b_l, b_u \in \mathbb{R}$.

Definition B.9 (*Monotone sequence*) A sequence $\{x_n\}$ is monotonic increasing or decreasing, if $x_{n+1} \geq x_n$ or $x_{n+1} \leq x_n$, respectively. It is strictly monotonic increasing or decreasing, if $x_{n+1} > x_n$ or $x_{n+1} < x_n$, respectively.

It should be noted, that every bounded monotone sequence converges but not every convergent sequence is monotone.

Differential calculus, for instance, is intimately related to the concept of infinite sequences, whereas integration relies on the idea of infinite sums also called series.

Definition B.10 (*Series*) The sum of infinitely many real numbers a_n related in a given way and listed in a given order is called a series written as $\sum_1^\infty a_n$.

The convergence of a series can be related to the convergence of sequences by means of the concept of partial sums.

Definition B.11 (*Partial sum of a series*) The n-th partial sum of a series $\sum_{i=1}^\infty a_i$ is defined by $S_n = \sum_{i=1}^n a_n$.

A series converges if the sequence of its partial sums converges to a limit. For finite sums, the order of summation does not affect at all the final result, e.g., $1 + 2 + 3 = 2 + 3 + 1 = 6$. Unfortunately, this remains true for series (infinite sums) only if they converge absolutely.

Definition B.12 (*Absolute convergence of a series*) A series $\sum_{i=1}^\infty a_i$ converges absolutely if $\sum_{i=1}^\infty |a_i|$ converges.

Sum and product of two absolute convergent series are again absolute convergent. The rules for computing the results coincide with the rules for finite sums.

B.3 Real Functions: Continuity and Boundedness

A real valued function of one real variable is a mapping $f : \Omega \to C$ with $\Omega, C \subseteq \mathbb{R}$. Ω and C are called domain and co-domain (or image). All definitions in this sections consider real valued functions of one real variable written as $f(x)$.

Definition B.13 (*Continuity of a function*) A function $f(x), x \in \Omega$ is continuous at $x = \hat{x}$ if for every real number $\varepsilon > 0$ there exists a $\delta > 0$ such that for every $y \in \Omega$ with $|y - \hat{x}| < \delta$ it follows that $|f(\hat{x}) - f(y)| < \varepsilon$.

Continuity according to Definition B.13 is also known as ordinary or point-wise continuity. It is a local property. A stronger form of continuity is uniform continuity which is a global property.

Definition B.14 (*Uniform continuity of a function*) A function $f(x), x \in \Omega$ is uniformly continuous if for every real number $\varepsilon > 0$ there exists a $\delta > 0$ such that for every $x, y \in \Omega$ with $|y - x| < \delta$ it follows that $|f(x) - f(y)| < \varepsilon$.

An even stronger continuity requirement is Lipschitz continuity which plays an important role in the theory of differential equations.

Definition B.15 (*Lipschitz continuity*) A function $f(x), x \in \Omega$ is Lipschitz continuous if for every $x, y \in \Omega$ there exists a positive real constant K such that $|f(x) - f(y)| \leq K|x - y|$.

Apart from continuity, boundedness is an important property of real functions.

Definition B.16 (*Boundedness*) A function $f(x), x \in \Omega$ is bounded from above if there exists a number K such that $f(x) \leq K$ for all $x \in \Omega$. It is bounded from below if there exists a number L such that $f(x) \geq L$ for all $x \in \Omega$.

The standard example $f(x) = \frac{1}{x}$ indicates that being bounded depends somehow on the properties of the domain, since f is bounded on the interval $[0.001, 1]$ but not on $(0, 1)$. Therefore, the question arises if there are conditions regarding the domain under which a continuous function is definitely bounded. At this point, the concept of compactness enters the scene. One possible line of argument to motivate compactness takes finite sets as a role model because real functions on finite sets are always bounded.

If A is a finite set, then a function $f : A \to \mathbb{R}$ is locally bounded, since it assigns to every element of A some real number. It is important to note that $\pm\infty$ do not belong to $\mathbb{R}$. One of these numbers will be the one with the largest absolute value, and f is bounded globally by the latter. This argument, trivial for finite sets, does not apply if A is infinite. Here, a similar line of argument is developed by means of the concept of open covers.

Definition B.17 (*Open cover*) A collection of open sets is a cover of a set A if A is a subset of the union of these open sets.

Suppose that an infinite set can be covered by a finite number of open sets. If, in addition f is bounded on all these open sets, the same argument used for finite sets can be adapted. However, there are many possible open covers for a given set and the finiteness argument must not depend on the choice of a particular cover. This gives raise to the following definition of compactness.

Definition B.18 (*Compactness*) A set A of a topological space is compact if **every** open cover of A has a finite sub-cover which covers A.

The open interval $(0, 1)$ **is not compact**

The open interval $I = (0, 1)$ can, of course, be covered by a finite number of open sets, starting with I itself. However, this does not mean that I is compact. In order to show that I is not compact, it suffices to find at least one open cover which does not have a finite sub-cover. The collection $O_n = (\frac{1}{n}, 1 - \frac{1}{n})$ covers, for $n \to \infty$, the open interval I but there is no finite n for which I is covered completely. Therefore, I is not compact.

Of course, to check compactness for a topological space or some subspace by means of Definition B.18 might not be trivial. However, it turns out that all closed intervals $[a, b]$ with $a, b \in \mathbb{R}$ are compact sets.

A specific subset of the domain of a function called support is frequently used, for instance, to make definitions more concise.

Definition B.19 (*Support of a real function*) Given a function $f : D \to \mathbb{R}$ with $D \subseteq \mathbb{R}$. The support of f, written as supp f, consists of all $x \in D$ with $f(x) \neq 0$.

B.4 Sequences and Series of Functions

The concepts of sequences and series can be generalized to functions defined on some interval. However, convergence becomes a more delicate issue giving raise to different criteria for convergence such as point-wise convergence and uniform convergence.

Definition B.20 (*Pointwise convergence of a sequence of functions*) A sequence of functions $\{f_n\}$, $f_n : \Omega \to \mathbb{R}$ defined on some domain $\Omega \subseteq \mathbb{R}$, converges pointwise to a limit function f if for any $\epsilon > 0$ and $x \in \Omega$ there exists a natural number N such that

$$|f_n = f| \leq \epsilon$$

for every $n > N$.

Definition B.21 (*Uniform convergence of a sequence of functions*) A sequence of functions $\{f_n\}$, $f_n : \Omega \to \mathbb{R}$ defined on some domain $\Omega \subseteq \mathbb{R}$, converges uniform to a limit function f if for any $\epsilon > 0$ there exists the same natural number N for every $x \in \Omega$ such that

$$|f_n = f| \le \epsilon$$

for every $n > N$.

For point-wise convergence N may depend on x whereas for uniform convergence one has to find an N independent on x. Point-wise convergence is a topological concept whereas the stronger concept of uniform convergence is not. Uniform convergence assures continuity of the limit function if the individual functions of a sequence are continuous whereas point-wise convergence does not.

B.5 Gradient and Differential in $\mathbb{R}$

The derivative of a function $f(x)$ at some $x = x_0$ is defined by

$$\left.\frac{\mathrm{d}f}{\mathrm{d}x}\right|_{x_0} = \lim_{h\to 0} \frac{1}{h}\left[f(x_0+h) - f(x_0)\right] . \tag{B.1}$$

If a unique limit exists, independent of the sign of h, $f(x)$ is continuous and continuously differentiable at $x = x_0$. The reverse does not hold, a fact that can be seen easily from the standard example $f(x) = |x|$, which is continuous at $x = 0$ but not continuously differentiable. The derivative at some arbitrary x is indicated by $\frac{\mathrm{d}f}{\mathrm{d}x}$, and performing this operation yields a function of x. That way, derivatives of arbitrary order can be defined for which the notation

$$\left.\frac{\mathrm{d}^k f}{\mathrm{d}x^k}\right|_{x_0} = \lim_{h\to 0} \frac{1}{h}\left[\frac{\mathrm{d}^{k-1}}{\mathrm{d}x^{k-1}} f(x_0+h) - \frac{\mathrm{d}^{k-1}}{\mathrm{d}x^{k-1}} f(x_0)\right] \quad k = 1, \dots, N$$

is commonly employed together with the definition

$$\frac{\mathrm{d}^0}{\mathrm{d}x^0} f := f .$$

This allows for a classification of functions by the following definition.

Definition B.22 (*Set of continuously differentiable functions*) The set of all functions defined on an interval $[a, b]$ or $[a, b)$, etc., having continuous derivatives up to order k is called $C^k([a, b])$, or $C^k([a, b))$, respectively, etc. If the interval is $(-\infty, +\infty)$, the notation $C^k(\mathbb{R})$ is used.

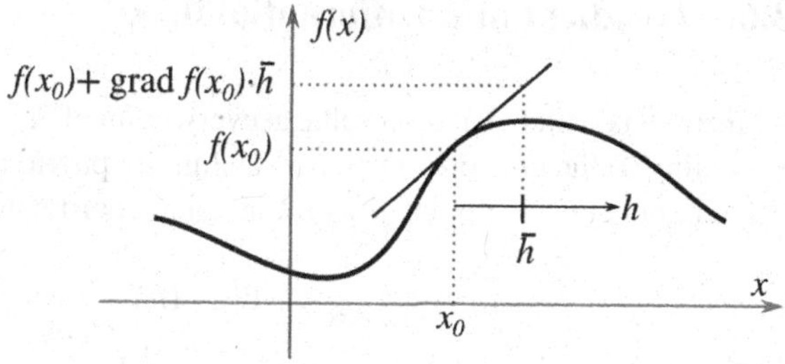

Fig. B.3 Illustration of gradient and differential in $\mathbb{R}$

It is not obvious how to transfer the concept of being continuously differentiable to higher dimensions. In order to obtain hints as to what a more general scheme might look like, it is useful to dissect (B.1) first. If f is continuously differentiable at x_0, then (B.1) exists and can be written as

$$\lim_{h\to 0} \frac{1}{h}\left[f(x_0+h) - f(x_0) - \left.\frac{\mathrm{d}f}{\mathrm{d}x}\right|_{x_0} h \right] = 0\,. \tag{B.2}$$

It reveals that being continuously differentiable implies the existence of a unique linear approximation of f at x_0,

$$\mathrm{d}|_{x_0} f(h) = \left.\frac{\mathrm{d}f}{\mathrm{d}x}\right|_{x_0} h = \operatorname{grad} f(x_0)\, h, \tag{B.3}$$

which is the differential of f at x_0. If f is continuously differentiable at x_0, then the differential is unique, because in this case, there exists, by definition, exactly one number $\operatorname{grad} f(x_0)$. In one dimension, this also implies linearity of (B.3). The differential (B.3) can also be written as

$$\operatorname{grad} f(x_0)\, h = \lim_{t\to 0} \frac{1}{t}\left[f(x_0+th) - f(x_0)\right], \tag{B.4}$$

which is the so-called directional derivative in one dimension. It has a straightforward generalization to higher dimensions. However, (B.4) must be handled with care, since it can also be computed for functions which are not continuously differentiable. The key is that continuous differentiability not only implies the existence of (B.4), but also its linearity wrt. h, or—vice versa—the existence of (B.4) implies continuous differentiability only if (B.4) is also linear wrt. h. Hence, the following statements are equivalent:

1. the function $f(x)$ is continuously differentiable at $x = x_0$,
2. there exists a unique linear approximation of $f(x)$ at $x = x_0$.

B.6 Gradient and Differential in $\mathbb{R}^N$

A function f, which defines a value at every point of $\mathbb{R}^N$ is called a scalar field in $\mathbb{R}^N$. According to the concept developed within the previous section, f is continuously differentiable at some point $\hat{p}$ with Cartesian coordinates $\hat{x}_k$ if a linear mapping

$$\mathrm{d}|_{\hat{p}}\, f(\boldsymbol{h})$$

exists analogously to (B.3). Linearity implies

$$\mathrm{d}|_{\hat{p}}\, f(\boldsymbol{h}) = h_i\, \mathrm{d}|_{\hat{p}}\, f(\boldsymbol{e}_i) = \left[\mathrm{d}|_{\hat{p}}\, f_i\, \boldsymbol{e}_i\right] \cdot \boldsymbol{h}$$

which reveals, that the gradient of a scalar field is a vector. To specify the components $\mathrm{d}|_{\hat{p}}\, f_i$ of this vector, we set $\boldsymbol{h} = \boldsymbol{e}_k$. In this case, we ask specifically for the change of f in direction of the coordinate x_k keeping all other coordinates fixed. But, this is precisely the definition of the partial derivative of f with respect to x_k at point $\hat{p}$, expressed concisely as

$$\left.\frac{\partial f}{\partial x_i}\right|_{\hat{p}} = \lim_{\varepsilon \to 0} \frac{1}{\varepsilon}[f(\hat{p} \dotplus \varepsilon\, \boldsymbol{e}_i) - f(\hat{p})]$$

by using hybrid addition according to Definition A.22. Therefore, the gradient of f at some point is given by

$$\operatorname{grad} f = \frac{\partial f}{\partial x_i} \boldsymbol{e}_i\,.$$

Applying this result for a scalar field in $\mathbb{R}^2$ yields

$$f(\hat{p} \dotplus \boldsymbol{h}) \approx f(\hat{p}) + \operatorname{grad} f(\hat{p}) \cdot \boldsymbol{h} = f(\hat{p}) + \left.\frac{\partial f}{\partial x_1}\right|_{\hat{p}} h_1 + \left.\frac{\partial f}{\partial x_2}\right|_{\hat{p}} h_2 \tag{B.5}$$

with $\hat{p} = (\hat{x}_1, \hat{x}_2)$ and $\hat{p} \dotplus \boldsymbol{h} = (\hat{x}_1 + h_1, \hat{x}_2 + h_2)$. It coincides with the result given in standard textbooks.

A function, which defines a vector at every point in some subset $\mathcal{B}$ of $\mathbb{R}^N$ is called a vector field in $\mathbb{R}^N$. Given a vector field $\boldsymbol{f}(x_1, x_2)$ in $\mathbb{R}^2$

$$\boldsymbol{f}(x_1, x_2) = f_1(x_1, x_2)\boldsymbol{e}_1 + f_2(x_1, x_2)\boldsymbol{e}_2\,.$$

Since the base vectors do not depend on location, the linear approximation affects only the functions f_1 and f_2. Therefore, the linear approximation of $\boldsymbol{f}$ at some point $\hat{p}$ in direction of a vector $\boldsymbol{h}$ is given by

$$
\boldsymbol{f}(\hat{p}+\boldsymbol{h}) \approx \boldsymbol{f}(\hat{p}) + \left[\left.\frac{\partial f_1}{\partial x_1}\right|_{\hat{p}} h_1 + \left.\frac{\partial f_1}{\partial x_2}\right|_{\hat{p}} h_2 \right] \boldsymbol{e}_1 + \left[\left.\frac{\partial f_2}{\partial x_1}\right|_{\hat{p}} h_1 + \left.\frac{\partial f_2}{\partial x_2}\right|_{\hat{p}} h_2 \right] \boldsymbol{e}_2
$$

where the two last terms on the right-hand side indicate the corresponding differential. Replacing h_1, h_2 by $\boldsymbol{e}_1 \cdot \boldsymbol{h}$ and $\boldsymbol{e}_2 \cdot \boldsymbol{h}$, respectively, yields the following result for the differential of a vector field at some arbitrary location

$$
\mathrm{d}\boldsymbol{f}(\boldsymbol{h}) = \left[\frac{\partial f_1}{\partial x_1} \boldsymbol{e}_1 \otimes \boldsymbol{e}_1 + \frac{\partial f_1}{\partial x_2} \boldsymbol{e}_1 \otimes \boldsymbol{e}_2 + \frac{\partial f_2}{\partial x_1} \boldsymbol{e}_2 \otimes \boldsymbol{e}_1 + \frac{\partial f_2}{\partial x_2} \boldsymbol{e}_2 \otimes \boldsymbol{e}_2 \right] \cdot \boldsymbol{h} \tag{B.6}
$$

The result (B.6) reveals, that the gradient of a vector field is a second order tensor. Furthermore, to express the differential symbolically as

$$
\mathrm{d}\boldsymbol{f} = \mathrm{grad}\, \boldsymbol{f} \cdot \boldsymbol{h}
$$

requires to define the gradient of a vector field as follows

$$
\mathrm{grad}\, \boldsymbol{f} := \frac{\partial f_k}{\partial x_i} \boldsymbol{e}_i \otimes \boldsymbol{e}_k \tag{B.7}
$$

in order to be consistent with the conventions established so far, see Sect. A.6. This scheme generalize straightforwardly to higher dimensions and Cartesian tensor fields of arbitrary order.

B.7 Notation Using Differential Operators in $\mathbb{R}^N$

A widely employed symbolic notation is based on the definition of the so-called Nabla-operator

$$
\nabla := \frac{\partial}{x_i} \boldsymbol{e}_i \, .
$$

Using $(\bullet)$ as a place holder, commonly required operations, such as gradient, divergence, and rotation, denoted by grad, div, and rot, can be defined as follows

$$
\mathrm{grad}\,(\bullet) := \nabla(\bullet) = \boldsymbol{e}_i \otimes \frac{\partial}{\partial x_i}(\bullet) \, , \tag{B.8}
$$

$$
\mathrm{div}\,(\bullet) := \nabla \cdot (\bullet) = \boldsymbol{e}_i \cdot \frac{\partial}{\partial x_i}(\bullet) \, , \tag{B.9}
$$

$$
\mathrm{rot}\,(\bullet) := \nabla \times (\bullet) = \boldsymbol{e}_i \times \frac{\partial}{\partial x_i}(\bullet) \, . \tag{B.10}
$$

The order of operands indicated explicitly by index notation matters. A number of different ways to define symbolic notation exist. Here, we follow [1].

The Poisson equation and its homogeneous version, the Laplace equation, are often expressed by means of the so-called Laplace operator, defined as

$$\Delta(\bullet) := \nabla \cdot \nabla(\bullet)\,. \tag{B.11}$$

A number of useful identities can be worked out, see, e.g., [1].

B.8 Notes on Riemann Integral and Lebesgue Integral

Given a real function $f(x)$ defined over some domain $\Omega \subseteq \mathbb{R}$. The graph of $f(x)$ is a curve defined by the set of points $(x, f(x))$. Interpreting x and $f(x)$ as Cartesian coordinates, the graph of $f(x)$ can be visualized in the $x-y$ plane with $y = f(x)$ as illustrated in Fig. B.4.

The definite integral of a function $f(x)$ is identified with the signed area of the region $\mathcal{B}$ between the graph of f and the x-axis. The sign is determined by f. Depending on $f(x)$, the region $\mathcal{B}$ can be a quite complicated geometric object whose exact signed area can be difficult to determine. However, one can at least try to estimate it using the following procedure:

1. Approximate $\mathcal{B}$ by primitive geometric objects for which assigning a signed area becomes a trivial task.
2. Estimate the area of $\mathcal{B}$ from the signed areas of the primitive objects used to approximate $\mathcal{B}$.

It is hoped that an approximation process based on this procedure converges eventually to the exact signed area of $\mathcal{B}$.

Although, engineers are usually familiar with the Riemann integral, it is briefly summarized here to prepare the reader for the Lebesgue integral which is the main subject of this section. To compute the Riemann integral of a real function defined over some domain $\Omega \subseteq \mathbb{R}$, a regular partitioning of Ω can be defined by dividing the domain into mutually disjoint sub-domains Ω_k.

Partitioning of $\Omega = [a, b)$

Given the interval $\Omega = [a, b)$. The collection of N half-open intervals $\Omega_k = [x_k, x_{k+1})$ with $x_k = a + [k-1]\frac{|b-a|}{N}$ defines a regular partitioning of Ω for some natural number N and $k = 1, \ldots, N$. A measure μ, more specifically a length, can be assigned trivially to every sub-domain by $\mu([x_k, x_{k+1})) = x_{k+1} - x_k$.
The acute reader will most likely ask what to do if $\Omega = [a, b]$. An answer to this question will be given in the context of the Lebesgue integral.

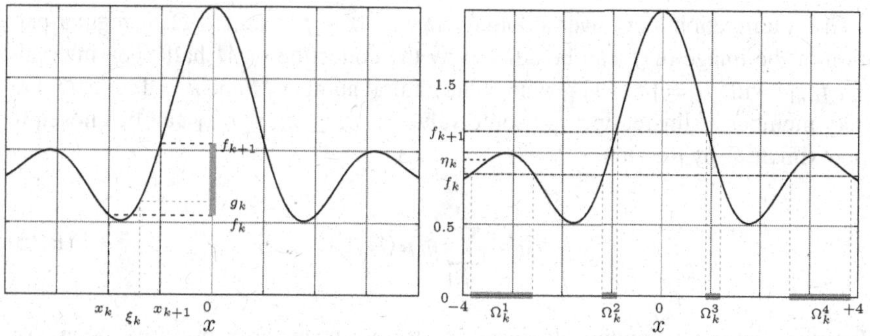

Fig. B.4 Riemann versus Lebesgue integration illustrated for $f(x) = \frac{\sin(x)}{x}\cos(3/2x) + 1$ in the interval $(-4, 4)$

Choosing some $\xi_k \in \Omega_k$, a rectangle with side lengths $\mu(\Omega_k)$ and $|f(\xi_k)|$ can be defined and the sum

$$s_N = \sum_{k=1}^{N} f(\xi_k)\,\mu(\Omega_k) \tag{B.12}$$

known as Riemann sum approximates the signed area of $\mathcal{B}$. The Riemann integral is now defined as the limit of the corresponding series

$$\int_{\Omega} f(x)\,\mathrm{d}x := \lim_{N\to\infty} s_N \tag{B.13}$$

provided, of course, such a limit exists. Recall, that ξ_k in (B.12) can be any point in Ω_k and, therefore, $f(\xi_k)$ is some representative of $f(x)$ in Ω_k. Considering the smallest and the largest value of $f(x)$ in Ω_k, given by L_k and U_k, respectively, two particular Riemann sums can be specified as follows

$$\underline{S}_N = \sum_{k=1}^{N} L_k\,\mu(\Omega_k)\ , \qquad \overline{S}_N = \sum_{k=1}^{N} U_k\,\mu(\Omega_k)$$

and (B.13) is only well defined if their limits exist and

$$\lim_{N\to\infty} \underline{S}_N = \lim_{N\to\infty} \overline{S}_N \tag{B.14}$$

holds.

To compute the Lebesgue integral, regular partitioning is applied to the image instead of the domain, as illustrated in Fig. B.4.

Given a function $f(x)$ over a domain Ω with $0 \leq f \leq d, x \in \Omega$. A regular partition of the image of f can be defined by the collection of M half-open intervals $[f_k, f_{k+1})$ with $f_k = [k-1]\frac{d}{M}$ with some natural number M and $k = 1, \ldots, N$. For every member of this collection, some value η_k, $f_k \leq \eta_k \leq f_{k+1}$ can be chosen to approximate $\mathcal{B}$ by the sum

$$\sigma_M = \sum_{j=1}^{M} \eta_k \mu(\Omega_k) . \tag{B.15}$$

However, the corresponding domains Ω_k are not necessarily primitive geometric objects which becomes even more obvious by considering the case of a function which depends on two real variables as illustrated in Fig. B.5.

To assign measures to rather general subsets of $\mathbb{R}^N$ requires a more rigorous approach known as Lebesgue measure theory which generalises the notion of volume. A detailed exposition of the subject is beyond the scope of this appendix but, following intuition and common sense, it is almost obvious what to expect from a measure μ:

- It should be a positive real number, i.e., given a set S and a measure μ, $\mu(S) \geq 0$ should hold.
- It should be invariant under translation and rotation.
- It should be additive. If a set S can be split into two disjoint sets S_1 and S_2 with measures $\mu(S_1)$, $\mu(S_2)$ then $\mu(S) = \mu(S_1) + \mu(S_2)$ should hold.
- It should be normed. For instance, the unit cube in $\mathbb{R}^3$ should have measure one.

A precursor of one of the fundamental ideas of Lebesgue measure theory has been mentioned already at the beginning of this section, namely the approximation of complicated geometric objects by primitive ones. Lebesgue measure theory generalises this idea by introducing explicitly the concept of Lebesgue measure, i.e., a collection of elementary sets in $\mathbb{R}^N$ to which a measure can be assigned easily. For instance, disjoint rectangles can be used for representing a more complicated set C in $\mathbb{R}^2$. According to additivity, mentioned above, the measure of C should then be obtained by summing up the measures of all rectangles required for representing C.

However, most sets can only be represented approximately by a finite number of elementary sets. Approximations of arbitrary precision can only be achieved by limit processes, which renders finite additivity mentioned above insufficient but countable additivity, also called σ-additivity, is required. More importantly, not all sets of $\mathbb{R}^N$ are actually Lebesgue measurable as shown by Vitali's theorem. Therefore, apart from making the notion of measure more precise, Lebesgue measure theory provides three crucial concepts

- measurable,
- sets of measure zero,
- almost everywhere.

A criterion to decide if a set is measurable requires some technicalities. The idea behind such a criterion, however, is relatively simple. Given, for instance, a set in terms of a region $\mathcal{B}$ in $\mathbb{R}^2$. A so-called inner measure $\underline{\mu}(\mathcal{B})$ requires, that all primitive objects for approximating $\mathcal{B}$ are inside $\mathcal{B}$. Analogously, an outer measure $\overline{\mu}(\mathcal{B})$ can be defined. With $\mu(\mathcal{B})$ denoting the exact measure of $\mathcal{B}$

$$\underline{\mu}(\mathcal{B}) \leq \mu(\mathcal{B}) \leq \overline{\mu}(\mathcal{B})$$

holds. The set $\mathcal{B}$ is measurable if the smallest possible outer measure and the largest possible inner measure converge to the same limit.

It can be shown, that sets of countably many points have measure zero by showing first, that the a single point $\hat{x}$ in $\mathbb{R}$ has measure zero. Consider the interval $\omega = (\hat{x} - \epsilon/2, \hat{x} + \epsilon/2)$. Defining the measure of an open interval (a, b) by $\mu((a, b)) = b - a$, implies $\mu(\omega) = \epsilon$. As ϵ approaches zero, ω and $\mu(\omega)$ approach $\{\hat{x}\}$ and zero, respectively. Hence, a single point has measure zero which implies

$$\mu((a, b)) = \mu([a, b]) = \mu([a, b)) = \mu((a, b]) = b - a\,.$$

Furthermore, a statement which is true, except for a subset of Ω with measure zero, is said to be true almost everywhere in Ω.

Conditions for the existence of the Lebesgue integral can be derived using the concept of simple functions by which it can be shown as well, that

$$\int_\Omega f(x)\,\mathrm{d}x = \lim_{M\to\infty} \sigma_M = \lim_{M\to\infty} \left\{ \sum_{j=1}^{M} \eta_j \mu(\Omega_j) \right\} \tag{B.16}$$

is a proper definition of the Lebesgue integral based on (B.15). However, this part is omitted here. Furthermore, computing the Lebesgue integral can become a laborious task due to the need for determining the domains for every particular (f_k, f_{k+1}). This becomes obvious when considering higher dimensional cases, see, for instance, Fig. B.5. The important message at this point is that if the proper Riemann integral exists the Lebesgue integral exists as well and both integrals coincide in value.

On the other hand, there are cases for which the Riemann integral does not exist but computing the Lebesgue integral is almost trivial as the following standard example shows.

Lebesgue integral of the Dirichlet function

Given the function

$$f(x) = \begin{cases} 0 & \text{if } x \text{ irrational} \\ 1 & \text{if } x \text{ rational} \end{cases}$$

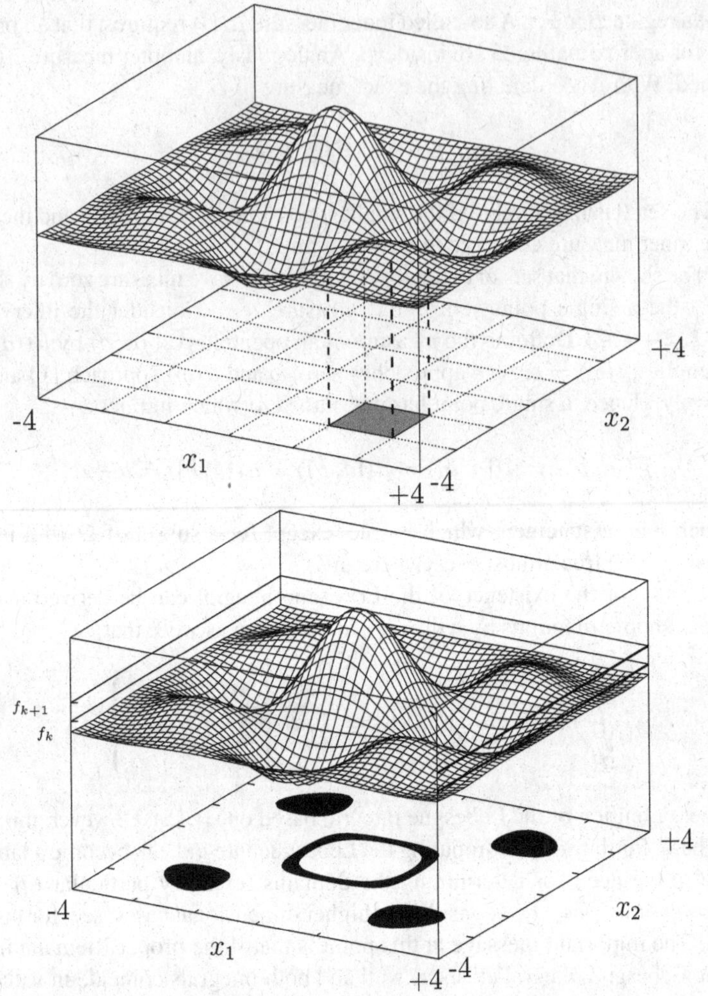

Fig. B.5 Illustration of Riemann (top) versus Lebesgue integration (bottom) in the interval $(-4, 4)$ for the function $f(x_1, x_2) = \frac{1}{\sqrt{x_1^2+x_2^2}}\left[\sin\left(\sqrt{x_1^2+x_2^2}\right)\cos\left(\frac{3}{2}x_1\right)\cos\left(\frac{3}{2}x_2\right)\right]+1$

in the domain $\Omega = (0, 1)$. Since there is only a countable number of rational numbers in Ω, the Lebesgue integral equals zero. A more detailed reasoning is left as an exercise to the reader.

The example illustrates, that the Lebesgue integral allows for a countable number of discontinuities whereas the Riemann integrals can only exist if the number of discontinuities is finite. Therefore, the Lebesgue integral applies to a broader class of functions. However, in most cases the advantage of the Lebesgue integral is conceptual rather than practical in that it allows for interchanging limiting operations under less strict conditions. Here, the conditions under which integration can be interchanged with limits of sequences of functions are of primary interest. For the Riemann integral,

$$\int_{\Omega} f(x)\,\mathrm{d}x = \int_{\Omega} \lim_{N\to\infty} \{f_k(x)\,\mathrm{d}x\}_{k=1}^{N}\,\mathrm{d}x = \lim_{N\to\infty} \left\{ \int_{\Omega} f_k(x)\,\mathrm{d}x \right\}_{k=1}^{N}$$

requires uniform convergence of the sequence $\{f_k(x)\}$ whereas point-wise convergence is sufficient in the case of the Lebesgue integral. Analogously,

$$\int_{\Omega} f(x)\,\mathrm{d}x = \lim_{N\to\infty} \sum_{k=1}^{N} \int_{\Omega} f_k(x)\,\mathrm{d}x$$

demands uniform convergence or just point-wise depending on whether the integral is a Riemann integral or a Lebesgue integral. For uniform convergence, see Definition B.21.

B.9 Comments on Notation for Integrals in $\mathbb{R}^N$

For the Riemann integral in $\mathbb{R}$ a standard notation has been established over time, see (B.13), and used in the same way in almost all texts regarding the matter. This is different for integration in $\mathbb{R}^N$. Intuitively, a notation like

$$\int\!\!\int_{\Omega} f(x_1, x_2)\,\mathrm{d}x_1\,\mathrm{d}x_2 = \int_{x_1=a_1}^{a_2} \int_{x_2=b_1}^{b_2} f(x_1, x_2)\,\mathrm{d}x_1\,\mathrm{d}x_2$$

seems to be the obvious choice for $\Omega \subset \mathbb{R}^2$, especially because the right hand side provides the details for an explicit computation. However, for general N, there are obvious disadvantages. Therefore, some authors prefer

$$\int_{\Omega} f\,\mathrm{d}^N x \tag{B.17}$$

instead, or a similar notation with minor variations. Such a notation is advantageous because it is short and flexible. It indicates unambiguously, that $\Omega \subset \mathbb{R}^N$ with coordinates $x_i, i = 1, .., N$. Therefore, this notation is also preferred by the author.

The integral theorems, provided in the next section, contrast integrals like (B.17) with corresponding integrals over the boundary $\partial\Omega$ of the considered domain Ω. A consistent use of the notation (B.17) is possible but there are some technical difficulties. For instance, if $\Omega \subset \mathbb{R}^3$ then its boundary $\partial\Omega$ is a surface in $\mathbb{R}^3$. Integration over general surfaces in $\mathbb{R}^3$ requires their parametrisation, i.e., a mapping into a chart $C \subset \mathbb{R}^2$ in which the integration is actually performed according to the properties of the parametrisation. In addition, a parametrisation is not unique.

In order to avoid such technicalities as long as specific computations are not necessary, the notation

$$\int_{\Omega} f \, \mathrm{d}^N x = \int_{\partial\Omega} g \, \mathrm{d}S$$

is preferred because of its independence of a particular parametrisation. In addition, the left hand side provides the necessary information about $\partial\Omega$, because $\Omega \subset \mathbb{R}^N$ implies, that integration over $\partial\Omega$ requires a parametrisation with respect to $\mathbb{R}^{N-1}$. Details, regarding the mentioned technicalities can be found in the main part of the book, more specifically in Sect. 2.3.5.

B.10 Integration Theorems in $\mathbb{R}^N$

A Lipschitz domain is a domain whose boundary coincides locally with a Lipschitz continuous function (see Definition B.15). A connected Lipschitz domain $\Omega \subset \mathbb{R}^N$ with boundary $\partial\Omega$ is considered. The Gauss-Integration theorem for a vector field $\boldsymbol{u}$ defined in Ω can be written using symbolic notation or index notation as follows

$$\int_{\Omega} \mathrm{div}(\boldsymbol{u}) \, \mathrm{d}^N x = \int_{\partial\Omega} \boldsymbol{u} \cdot \boldsymbol{n} \, \mathrm{d}S \qquad \int_{\Omega} u_{i,i} \, \mathrm{d}^N x = \int_{\partial\Omega} u_i n_i \, \mathrm{d}S \tag{B.18}$$

where $\boldsymbol{n}$ indicates the unit normal vector at $\partial\Omega$. (B.18) is also known as Gauss divergence theorem in $\mathbb{R}^N$. Setting $\boldsymbol{u} = w\boldsymbol{c}$, where w is a scalar field and $\boldsymbol{c}$ is a constant vector, the Gauss gradient theorem

$$\int_{\Omega} \mathrm{grad}(w) \, \mathrm{d}^N x = \int_{\partial\Omega} w\boldsymbol{n} \, \mathrm{d}S \qquad \int_{\Omega} w_{,j} \, \mathrm{d}^N x = \int_{\partial\Omega} w \, n_j \, \mathrm{d}S \tag{B.19}$$

is derived.

These integration theorems are valid for Riemann integrable functions. Corresponding Lebesgue counterparts exist but require substantially more theoretical effort.

B.11 Mappings and Their Jacobians

Using reference domains for FEM solution schemes implies to work with the same functions in different coordinate systems (coordinate charts) which are related by corresponding mappings. These mappings are in general nonlinear. Affine mappings, i.e., linear mappings with an offset, are actually special cases.

Nonlinear chart relations affect differentiation and integration which is illustrated first for the one-dimensional examples sketched in Fig. B.6.

Given the two domains $\Omega_\square$ and Ω with coordinate systems indicated by the coordinates ξ and x, respectively. $\Omega_\square$ and Ω are related via a mapping χ given by

$$\chi : \Omega_\square \rightarrow \Omega$$
$$\xi \mapsto \frac{1}{2}[1+\xi]^2$$

meaning that for a given point in $\Omega_\square$ with coordinate ξ, its image in Ω is addressed by the coordinate $x = \frac{1}{2}[1+\xi]^2$.

In addition, the function $\overline{f}(x) = [x+1]^2 + 1$ is considered for $\Omega = (0, 2)$. The representation of $\overline{f}$ in $\Omega_\square$ is obtained by substituting the coordinate x in $\overline{f}$ according to χ which yields

$$f(\xi) = \left[\frac{1}{2}[1+\xi]^2 + 1\right]^2 + 1\,.$$

This operation and its inverse are usually expressed as

$$f(\xi) = \overline{f} \circ \chi \qquad\qquad \overline{f}(x) = f \circ \chi^{-1}\,. \tag{B.20}$$

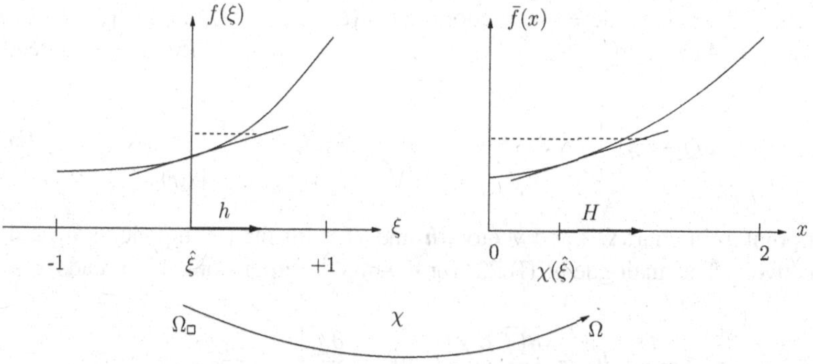

Fig. B.6 Example for a function given in two different coordinate systems (charts) which are related via a nonlinear mapping

where χ^{-1} is obtained by inverting χ accordingly. For the example considered here the result is $\xi = \sqrt{2x} - 1$.

Given a point in $\Omega_\square$ with coordinate $\hat{\xi}$. Its coordinate in Ω is given by $x = \chi(\hat{\xi})$. Differentials referring to the same point can be computed in both charts according to (B.3) as follows

$$\mathrm{d}|_{\hat{\xi}} f(h) = \left.\frac{\partial f}{\partial \xi}\right|_{\hat{\xi}} h\,, \qquad \mathrm{d}|_{\chi(\hat{\xi})} \overline{f}(H) = \left.\frac{\partial \overline{f}}{\partial x}\right|_{\chi(\hat{\xi})} H \tag{B.21}$$

and the results depend on the values h and H. Equal results can only be obtained if h and H are related according to the correspondence between $\Omega_\square$ and Ω given by χ. To find this relation, chain rule is applied after using (B.20) in the left part of (B.21) as follows

$$\left.\frac{\partial f}{\partial \xi}\right|_{\hat{\xi}} h = \left.\frac{\partial (\overline{f} \circ \chi)}{\partial \xi}\right|_{\hat{\xi}} h = \left.\frac{\partial \overline{f}}{\partial x}\right|_{\chi(\hat{\xi}} \left.\frac{\partial \chi}{\partial \xi}\right|_{\hat{\xi}} h\,. \tag{B.22}$$

Comparing the result with the corresponding part of (B.21) reveals, that equal results for the differentials can only be obtained if the following linear relation between h and H holds

$$H = \left.\frac{\partial \chi}{\partial \xi}\right|_{\hat{\xi}} h\,.$$

The term $\left.\frac{\partial \chi}{\partial \xi}\right|_{\hat{\xi}}$ is the linearisation of the mapping χ at a given point with coordinate $\hat{\xi}$, known as the Jacobian of χ at $\hat{\xi}$. Analogously, the Jacobian of χ^{-1} can be determined which is left as an exercise to the reader.

This scheme extends analogously to higher dimensions. Two charts $\Omega_\square \subset \mathbb{R}^N$ and $\Omega \subset \mathbb{R}^N$ are considered with coordinates $(\xi_1, \xi_2, \ldots, \xi_n)$ and $(x_1, x_2, \ldots, x_n)$, respectively. According to Sect. B.6, the differentials in the different charts read

$$\mathrm{d}|_p f(\boldsymbol{h}) = \left.\frac{\partial f}{\partial \xi_i}\right|_p h_i\,, \qquad \mathrm{d}|_{\chi(\hat{\xi})} \overline{f}(\boldsymbol{H}) = \left.\frac{\partial \overline{f}}{\partial x_i}\right|_{\chi(p)} H_i \tag{B.23}$$

for a point p in chart $\Omega_\square$ and vectors $\boldsymbol{h}$ and $\boldsymbol{H}$ with their components h_i and H_i, respectively. The analogue of (B.22) for the higher dimensional case reads

$$\left.\frac{\partial f}{\partial \xi_k}\right|_p h_k = \left.\frac{\partial (\overline{f} \circ \chi)}{\partial \xi_k}\right|_p h_k = \left.\frac{\partial \overline{f}}{\partial x_i}\right|_{\chi(p)} \left.\frac{\partial \chi_i}{\partial \xi_k}\right|_p h_k\,,$$

from which the Jacobian of χ can be deduced by comparison with (B.23). Analogously, the Jacobian of χ^{-1} can be determined. The final results are the second order tensors

$$\mathsf{F} = \frac{\partial \chi_k}{\partial \xi_i} \boldsymbol{e}_i \otimes \boldsymbol{e}_k\,, \qquad \mathsf{F}^{-1} = \frac{\partial \chi_k^{-1}}{\partial x_i} \boldsymbol{e}_i \otimes \boldsymbol{e}_k\,.$$

B.12 Bibliographical Remarks

For readers interested in Point Set Topology, the standard book [2] is a good start. An insightful didactic text about the subject is surely [3].

Concepts of Real Analysis are laid out in a didactic and comprehensible manner in [4]. Equally recommendable is [5]. A detailed and insightful discussion about Lebegue integration is provided by [6].

References

1. G. Holzapfel, *Nonlinear Solid Mechanics: A Continuum Approach for Engineering* (Wiley, New York, 2000)
2. B. Mendelson, *Introduction to Topology*, 3rd edn. Dover Books on Mathematics (Dover Publications, 2012)
3. R. Geroch, *Topology* (Minkowski Institute Press, 2013)
4. S. Abbott, *Understanding Analysis* (Springer, Berlin, 2015)
5. B.S. Thomson, J.B. Bruckner, A.M. Bruckner, *Elementary Real Analysis*, vol. 1, 2nd edn. (Prentice-Hall, London, England, 2008)
6. R.E. Wernikoff, *Outline of Lebesgue Theory: A Heuristic Introduction*. Technical Report. Research Laboratory of Electronics (MIT, 1957)

Appendix C
Elements of Linear Functional Analysis

Abstract The chapter provides a minimum of information required to understand the logical structure of Functional Analysis as well as its importance for FEM. Function spaces are introduced as a generalization of the idea of finite dimensional normed real vector spaces. Banach space and Hilbert space are defined before introducing ordinary weak derivatives and the concept of Sobolev spaces. The Lax-Milgram lemma is discussed and its application to linear boundary value problems is illustrated. For the sake of comprehensibility, only boundary value problems whose strong forms are given by ordinary differential equations are considered in this chapter, except for the last section, which provides essential generalisations required in the context of partial differential equations.

C.1 Motivation

Sytems of linear algebraic equations represent mappings between finite dimensional vector spaces. Consider the generic notation

$$T\,x = b$$

in terms of an operator T acting on vectors, for instance $x \in \mathbb{R}^N$. The properties of T determine if a unique solution exists. A similar generic notation can be used for linear differential or integral equations

$$L\,u = f$$

with a differential or integral operator L, for instance $L = \frac{\mathrm{d}}{\mathrm{d}x}$. L acts on a function $u : \Omega \to \mathbb{R}$ defined on some domain $\Omega \subset \mathbb{R}$ or even $\Omega \subseteq \mathbb{R}$.

Since summation of functions and multiplication of functions with a real number are well defined, the idea of vector spaces the elements of which are functions emerges

U. Mühlich, *Enhanced Introduction to Finite Elements for Engineers*, Solid Mechanics and Its Applications 268, https://doi.org/10.1007/978-3-031-30422-4

rather naturally together with the expectation, that concepts from Linear Algebra apply as well for linear differential equations, i.e., it is hoped, that linear differential or integral equations can be dealt with in pretty much the same way as linear algebraic equations. This is where Linear Functional Analysis departs.

The space of all third order polynomials defined on a given domain with summation and multiplication by a real numbers is a vector space. Its elements are functions. Moreover, it is isomorphic to $\mathbb{R}^3$. However, such spaces are in general not suitable for solving differential equations. As an example for a more suitable approch, recall the representation of a periodic function $f(x)$ by means of a Fourier cosine series

$$f(x) = a_0 + \sum_{j=1}^{\infty} a_j \cos(jx), \quad x \in \Omega. \tag{C.1}$$

Apparently, the function $f(x)$ is expressed in terms of a basis given by $\{1, \cos(x), \cos(2x), \ldots\}$ of infinite dimension. This indicates, that vector spaces of Functional Analysis, also called function spaces, are infinite dimensional. Therefore, it can not be expected, that everything from Linear Algebra and Real Analysis carries over smoothly to Functional Analysis. Hence, gaining a certain understanding of the subject implies a considerable amount of additional work but the benefits are usually worth it. For instance, in view of the well-known Gibbs-phenomenon, i.e., the overshot behaviour of Fourier-series at jump discontinuities, the interpretation of the equality sign in (C.1) is not really clear without further information. Functional Analysis provides answers to questions like this.

Eventually, Linear Functional Analysis can be summarized as analysis in infinite dimensional vector spaces. Therefore, the following exposition relies heavily on the content of Appendices A and B.

C.2 Introduction to Function Spaces

The very first task consists in defining vector spaces the elements of which are functions over some interval $\Omega \subseteq \mathbb{R}$. In view of (C.1), infinite dimensional spaces should be considered. As for finite dimensional vector spaces, analysis requires at least a topology. Normed vector spaces are of particular interest because a norm induces a metric which in turn induces a topology.

Furthermore, completeness is required for working safely with limit processes and the concept of Cauchy-sequences defined in Definition B.3 can be generalized for normed vector spaces as follows.

Definition C.1 (*Cauchy-sequences in normed vector spaces*) Let $\mathcal{V}$ be a vector space with norm $||v||$, $v \in \mathcal{V}$. A sequence $\{u_n\}$ is a Cauchy sequence if there is a positive integer N such that

$$||u_{n+1} - u_n|| \leq \epsilon$$

for all $n \geq N$ with $\epsilon \in \mathbb{R}$.

The reasoning so far leads to a rather general definition for a space, known as Banach space

Definition C.2 (*Banach space*) A Banach space is a normed vector space which is complete under the defined norm.

Definition C.2 provides a basic setting for performing analysis safely. A general theme in Functional Analysis is to ensure that specific function spaces are at least Banach spaces. All finite dimensional normed vector spaces are Banach spaces.

The space of all k-times continuously differentiable functions defined for domain Ω, for instance the open interval $\Omega = (0, 1)$ is of particular interest.

Definition C.3 (*The space* $C^k(\Omega)$) A function is $k-$times continuously differentiable in the domain Ω, if its k-th derivative is a continuous function. The set of all $k-$times continuously differentiable functions in Ω is called $C^k(\Omega)$.

Remark C.1 $C^k(\Omega)$ is a vector space because summation of $C^k(\Omega)$ functions as well as their multiplication with real numbers are well defined. Functions which belong to $C^\infty(\Omega)$ are called smooth functions.

The question arises whether or not $C^k(\Omega)$ can be made a Banach space by defining a norm under which this space is complete. The properties of a norm have been defined already in Appendix A, see Definition A.15. Since a norm assigns a positive real number to an element of a vector space, the integral of the absolute value of a function over the domain Ω seems to be a suitable candidate, e.g.,

$$||f|| := \int_\Omega |f(x)| \, \mathrm{d}x \, . \tag{C.2}$$

The metric $||f - g||$ induced by such a norm provides a number which reflects the difference between two functions f and g regarding the entire domain.

However, whether or not the integral exists for every member of $C^k(\Omega)$ depends on Ω. In addition, the limit function of a Cauchy-sequence of Riemann-integrable functions is not necessarily Riemann-integrable. Therefore, the space of all Riemann-integrable functions defined over a domain Ω is not complete which renders such spaces useless for Functional Analysis. The Lebesgue-integral discussed in Appendix B is more suitable for ensuring completeness which is the reason for defining eventually the so-called L^p-norm.

Definition C.4 (*L^p-norm*) Given a Lebesgue integrable function defined on a domain Ω. The norm

$$||f||_{L^p} = \left[\int_{\Omega} [f(x)]^p \mathrm{d}x \right]^{1/p}$$

is called L^p-norm with $p \in \mathbb{N}$.

According to Definition A.15, the triangle inequality is a defining property of a norm. This is ensured for L^p-norms by the so-called Minowski inequality

$$||f + g||_{L^p} \leq ||f||_{L^p} + ||g||_{L^p} \,. \tag{C.3}$$

Definition C.5 (*L^p space*) The space of all functions for which the integral defined in Definition C.4 exists, i.e.,

$$\left[\int_{\Omega} [f(x)]^p \mathrm{d}x \right]^{1/p} < \infty$$

is called $L^p(\Omega)$ with $p \in \mathbb{N}$.

As in Linear Algebra, an inner product induces a norm and inner product spaces play an important role as well in Functional Analysis. The completeness issue however, requires an intermediate step. For the formal definition of an inner product see Definition A.16.

Definition C.6 (*Pre-Hilbert space*) A vector space with inner product is a Pre-Hilbert space.

Definition C.7 (*Hilbert space*) A Hilbert space is a vector space with inner product which is complete with respect to the induced norm.

It is easy to show, that the L^2 norm can be induced by the following inner product

$$(f, g) := \int_{\Omega} f(x)g(x)\,\mathrm{d}x$$

which is left as an exercise to the reader. Hence, L^2 spaces are Hilbert spaces. In the context of function spaces, the notation (f, g) is preferred instead of $f \cdot g$. The Cauchy-Schwarz inequality for L^2 spaces reads

$$|(f, g)| \leq ||f||_{L^2} ||g||_{L^2} \tag{C.4}$$

which is a special case of the so-called Hölder's inequality

$$||fg||_{L^1} \leq ||f||_{L^p}||g||_{L^q} \tag{C.5}$$

for $p, q \in \mathbb{N}$ and $\frac{1}{p} + \frac{1}{q} = 1$.

C.3 Linear Mappings and Linear Forms

As in Linear Algebra, linear mappings, often called linear operators, between spaces are of utmost importance in Linear Functional Analysis. Definition A.24 is rewritten for normed vector spaces as follows.

Definition C.8 (*Linear mapping between normed vector spaces*) Given two normed vector spaces $\mathcal{V}$ and $\mathcal{W}$. A mapping $T : \mathcal{V} \to \mathcal{W}$ is linear if

$$T(\alpha v + \beta w) = \alpha T(v) + \beta T(w)$$

with $v \in \mathcal{V}$, $w \in \mathcal{W}$ and $\alpha, \beta \in \mathbb{R}$.

The most frequent linear mappings in the context of Finite Elements are linear and bilinear forms.

Definition C.9 (*Linear form*) Given a normed vector space $\mathcal{V}$. A linear form F is a linear mapping $F : \mathcal{V} \to \mathbb{R}$.

Definition C.10 (*Bilinear form*) Given a normed vector space $\mathcal{V}$. A bilinear form F is a mapping $F : \mathcal{V} \times \mathcal{V} \to \mathbb{R}$ which is linear in both arguments.

Linear mappings between finite dimensional vector spaces are continuous. Mappings between infinite dimensional vector spaces are continuous if and only if they are linear and bounded.

Definition C.11 (*Bounded mappings*) Given two normed vector spaces $\mathcal{V}$ and $\mathcal{W}$. A mapping $T : \mathcal{V} \to \mathcal{W}$ is bounded if there exists a positive number K such that

$$||T(v)||_{\mathcal{W}} \leq K||v||_{\mathcal{V}}$$

holds for all $v \in \mathcal{V}$. A mapping $T : \mathcal{V} \times \mathcal{V} \to \mathcal{W}$ is bounded if there exists a positive number K such that

$$||T(u, v)||_{\mathcal{W}} \leq K||u||_{\mathcal{V}}||v||_{\mathcal{V}}$$

holds for all $u, v \in \mathcal{V}$.

In the context of the Lax-Milgram lemma to be discussed later, coercive bilinear forms play an essential role.

Definition C.12 (*Coercive bilinear form*) Given a Hilbert space $\mathcal{V}$. A bilinear form $A : \mathcal{V} \times \mathcal{V} \rightarrow \mathbb{R}$ is coercive if there exists a positive number K such that

$$A(v, v) \geq K||v||^2_{\mathcal{V}}$$

holds for all $v \in \mathcal{V}$.

C.4 Weak Derivative

The weak derivative is motivated by the well-known integration by parts theorem. Its precise definition requires some preliminary work. The closure of a set A, written as $\text{cl}(A)$ can be defined as the union of the set itself and its boundary ∂A, i.e. $\text{cl}(A) = A \bigcup \partial A$. A suitable test function space can be defined concisely in terms of concepts introduced by the definitions Definitions C.3 and B.19 together with the concept of closure.

Definition C.13 $[C_c^\infty(\Omega)]$ The space of all functions $v \in C^\infty(\Omega)$ for which $\text{cl}(\text{supp}\, v)$ is compact is called $C_c^\infty(\Omega)$.

The weak derivative can now be defined for real functions of one real variable as follows.

Definition C.14 (*Weak derivative in* $\mathbb{R}$) Given a Lebesgue integrable function f on $\Omega = (a, b)$. If

$$\int_a^b g\, v\, \mathrm{d}x = (-1)^k \int_a^b f \frac{\mathrm{d}^k}{\mathrm{d}x^k} v\, \mathrm{d}x$$

exists for arbitrary test functions $v \in C_c^\infty(\Omega)$, the function g is called the k−th weak derivative of f in Ω and denoted by $D^k f$ with $D^0 f := f$.

Remark C.2 Test functions must vanish at the boundary of Ω to avoid boundary terms which could be achieved as well by a less restrictive test function space. However, since there are no restrictions for f, the test function space must assure that Definition C.14 provides a safe tool to work with. More specifically, using test functions from $C_c^\infty(\Omega)$ ensures, that the existence of the integrals in Definition C.14 can only be jeopardized by f but not by any test function.

Existence of a strong derivative implies the existence of its weak counterpart simply by the integration by parts theorem. The concept of weak derivatives can be generalized for $\mathbb{R}^N$ in terms of weak partial derivatives, which is discussed in Sect. C.9.

C.5 Sobolev Spaces

Sobolov spaces are normed function spaces with L^p norms including also weak derivatives. In the relevant literature they are often denoted by $W^{k,p}$ where k indicates the order of the highest derivative and p refers to the L^p norm. Here, only the case $p = 2$ is considered.

Definition C.15 (*The space* $W^{k,2}(\Omega)$) The space of all real functions f over Ω with finite L^2 norm

$$||f||_{1,2} = \left(\sum_{i=0}^{1} \int_{\Omega} (D^i f)^2 \, \mathrm{d}x \right)^{\frac{1}{2}}$$

is called the Sobolev space $W^{k,2}(\Omega)$.

Since the L^2 norm corresponds to an inner product, all $W^{k,2}(\Omega)$ spaces are Hilbert spaces. Therefore, the space $W^{k,2}(\Omega)$ is often just called $H^k(\Omega)$.

Hilbert space property of $W^{1,2}(\Omega)$

The space $W^{1,2}(\Omega) = H^1(\Omega)$ is a Hilbert space with inner product

$$(f, g) = \int_{\Omega} f \, g \, \mathrm{d}x + \int_{\Omega} D^1 f \, D^1 g \, \mathrm{d}x$$

which illustrates the relation between $W^{k,2}(\Omega)$ spaces and weak forms of boundary value problems. The reader is invited to consider $f'' - f + n = 0$ with homogeneous Dirichlet boundary conditions, interpreting all derivatives in the variational form as weak derivatives.

C.6 Variational Formulation of Boundary Value Problems

Within this section, the boundary value problem given by

$$u'' + g = 0 \quad x \in \Omega = (a, b) \tag{C.6}$$

is considered for illustration purposes together with different boundary conditions. The prime is used as short hand notation for the classical derivative with respect to the spatial coordinate. The task of solving the boundary value problem can be stated as follows. Given $g \in C^2(\Omega)$. Find $u \in C^2(\Omega)$ such that (C.6) is fulfilled for every $x \in \Omega$ and given boundary conditions.

Multiplying (C.6) with a function $v \in C^1(\Omega)$ and applying integration by parts yields eventually

$$\int_a^b u'v' \,\mathrm{d}x - \int_a^b gv \,\mathrm{d}x + u(0)v(0) - u(l)v(l) = 0\,. \tag{C.7}$$

First, homogeneous Dirichlet boundary conditions, i.e., $u(a) = 0$ and $u(b) = 0$, are considered. Setting $v(a) = 0$ and $v(b) = 0$ yields the variational formulation

$$\int_a^b u'v' \,\mathrm{d}x - \int_a^b gv \,\mathrm{d}x = 0 \tag{C.8}$$

$$u(a) = 0$$
$$u(b) = 0$$

It is an easy exercise to show, that this variational formulation is equivalent to (C.6) with homogeneous Dirichlet boundary conditions, provided that v is bounded and sufficiently regular. It only requires to apply integration by parts backwards and is left as an exercise to the reader. Equivalence means, that a function u which solves (C.6) for homogeneous Dirichlet boundary conditions solves (C.8) and vice versa.

For mixed boundary conditions, for instance, $u(a) = 0$ and $u'(b) = P_b$, the variational formulation reads

$$\int_a^b u'v' \,\mathrm{d}x - \int_a^b gv \,\mathrm{d}x - P_b v(l) = 0 \tag{C.9}$$

$$u(a) = 0$$

which, again, is equivalent to (C.6) with the same mixed boundary conditions, provided that v and g are bounded and sufficiently regular.

However, the variational formulations (C.8) and (C.9) make sense as well if they are considered as independent of their corresponding strong forms. In this case, it is sufficient to demand that $u, v \in C^1(\Omega)$ and $g \in C^0(\Omega)$ apart from being bounded.

To find a weak form or, more specifically, weak solutions of a boundary problem, it is asked how much the continuity requirements can be weakened such that the variational formulation still makes sense and classical solutions are still included as special cases. The first question which comes to mind is whether or not u' and v' in (C.8) and (C.9) must be classical derivatives and the answer is no. The variational forms can still make sense as tasks on their own if u' and v' are interpreted as weak derivatives, by which Sobolov spaces become an appropriate setting for solutions.

However, the use of Sobolov spaces introduces new challenges regarding the realization of boundary conditions. This is discussed in more detail together with the problem of existence and uniqueness of solutions within the next section.

C.7 The Lax-Milgram Lemma

The Lax-Milgram lemma considers a bilinear functional $a : \mathcal{V} \times \mathcal{V} \to \mathbb{R}$ on a Hilbert space $\mathcal{V}$. It states, that if $a(u, \phi)$ is continuous and coercive then for every continuous linear form $b : \mathcal{V} \to \mathbb{R}$ there exists a unique $u \in \mathcal{V}$ such that

$$a(u, \phi) = b(\phi) \tag{C.10}$$

holds for all $\phi \in \mathcal{V}$.

The variational form of every linear boundary value problem can be cast into the form given by (C.10). Therefore, the question whether or not a unique solution exists for a given boundary value problem can be answered by applying the criteria involved in the Lax-Milgram lemma after choosing an appropriate space $\mathcal{V}$.

Operator notation for (C.9)

Regarding the variational form (C.9), $a(u, v)$ and $b(v)$ are given by

$$a(u, v) = \int_{\Omega} u'\, v'\, \mathrm{d}x \quad , \qquad b(v) = \int_{\Omega} g\, v\, \mathrm{d}x + P_b v(l)\,.$$

In the following, only $H^1(\Omega)$ is considered for illustration purposes. In order to account for homogeneous Dirichlet boundary conditions the space $H_0^1(\Omega)$ is defined.

Definition C.16 (*The space* $H_0^1(\Omega)$) The space $H_0^1(\Omega)$ is the completion of $C_c^\infty(\Omega)$ under the norm of $H_1(\Omega)$.

Definition C.16 uses of the fact that $C_c^\infty(\Omega)$ is dense in $H^1(\Omega)$ in the sense of Definition B.7, It means, that limits of sequences of elements of $C_c^\infty(\Omega)$ can be used to represent the elements of $H^1(\Omega)$ and the norm of $H_1(\Omega)$ recognizes these limits as proper members of $H_1(\Omega)$. As a result, all members of $H_0^1(\Omega)$ vanish automatically at $\partial\Omega$. In addition, for $u \in H_0^1(\Omega)$, Poincaré's inequality

$$||u||_{L^2(\Omega)} \leq K||u'||_{L^2(\Omega)} \tag{C.11}$$

holds with a positive constant K.

C.8 Weak Solutions of Boundary Value Problems

Problem (C.8) is considered for illustration purposes. Elements of $H_0^1(\Omega)$ are suitable candidates for a solution because of the highest order of derivatives in $a(u, v)$ and the homogeneous Dirichlet boundary conditions.

The Lax-Milgram lemma can be applied only if $H_0^1(\Omega)$ is also used as test function space. This is only possible if the integrals remain well defined. Hence, the function g must fulfil certain requirements, see, as well, Remark C.2.

The task of finding a weak solution of (C.8) can be expressed as follows.

Task C.1 Given $\Omega = (a, b)$ and $g \in L^2(\Omega)$. Find $u \in H_0^1(\Omega)$ such that

$$\int_a^b u' \, v' \mathrm{d}x - \int_a^b g \, v \, \mathrm{d}x = 0 .$$

holds for arbitrary $v \in H_0^1(\Omega)$.

For a unique solution to exist, $a(u, v)$ has to be continuous and coercive. To be continuous, it has to be linear and bounded. Since it is easy to check, that $a(u, v)$ and $b(v)$ are bilinear and linear, respectively, with respect to their arguments, this task is left as an exercise to the reader. It remains to show that $a(u, v)$ and $b(u)$ are bounded. In addition $a(u, u)$ must be coercive. The corresponding proofs can be found in many textbooks. They are partially repeated here for demonstrating a general pattern in Functional Analysis, namely the use of estimates and inequalities.

Since $a(u, v)$ maps from $H^1(\Omega) \times H^1(\Omega)$ to $\mathbb{R}$, checking boundedness according to Definition C.11 involves the following steps

$$|a(u, v)| = \left| \int_a^b u' \, v' \mathrm{d}x \right| \leq \int_a^b |u' \, v'| \mathrm{d}x = ||u'v'||_{L^1(\Omega)} \tag{C.12}$$

$$\leq ||u'||_{L^2(\Omega)} ||v'||_{L^2(\Omega)} \tag{C.13}$$

$$\leq ||u||_{H^1(\Omega)} ||v||_{H^1(\Omega)} \tag{C.14}$$

which will be explained in more detail next. Moving the absolute value operation under the integral sign produces a larger or equal value since integration is a summation process. The result is just the L^1 norm in Ω, see Definition C.4. Step (C.13) applies Hölder's inequality (C.5). The last step is based on the fact, that

$$||w'||_{L^2(\Omega)} = \sqrt{\int_\Omega |w'|^2 \,\mathrm{d}x} \le \sqrt{\int_\Omega |w|^2 \,\mathrm{d}x + \int_\Omega |w'|^2 \,\mathrm{d}x} = ||w||_{H^1(\Omega)}$$

holds for $w \in H^1(\Omega)$. Using a similar line of arguments, it can be shown, that $a(u, u)$ is coercive according to Definition C.12. It requires Poincaré's inequality (C.11). This step, however, is omitted here. Eventually, it has to be checked whether or not $b(v)$ is bounded which yields

$$|b(v)| \le K||v||_{H^1(\Omega)}$$

with $K = ||g||_{L^2(\Omega)}$. It shows, that continuity of $b(v)$ requires g to be bounded with respect to the L^2 norm.

Remark C.3 (*Regularity*) According to the Sobolev embedding theorem, the solution of Task 1 is continuous. Since $g \in L^2(\Omega)$, it can have a finite number of discontinuities. If $g \in C^1(\Omega)$, the classical solution is obtained, i.e., $u \in C^2(\Omega)$.

C.9 Generalisations

So far, essential aspects of Functional Analysis in the context of FEM have been discussed only for boundary value problems whose strong forms are ordinary differential equations. Extending the approach to partial differential equations requires to generalise a number of concepts.

This affects in particular Definitions C.3, C.4, C.5, and C.13. Fortunately, the corresponding generalisations are available and the necessary modifications are rather straight forward. Firstly, $\Omega \subset \mathbb{R}^N$ has to be considered instead of $\Omega \subset \mathbb{R}$. Secondly, ordinary derivatives have to be replaced by partial derivatives, and, lastly, '$\mathrm{d}x$' has to be substituted by '$\mathrm{d}^N x$' in all integrals.

Repeating all definitions mentioned above in their generalised form does not provide any additional benefit for the learning process. Therefore, it is omitted here with the understanding, that for what follows, the corresponding modifications mentioned above apply.

As already mentioned, the concept of weak differentiability can be extended to partial derivatives.

Definition C.17 (*Weak first-order partial derivative in* $\mathbb{R}^N$) Given a Lebesgue integrable function f on $\Omega \subset \mathbb{R}^N$. If

$$\int_\Omega g_i \, v \,\mathrm{d}^N x = -\int_\Omega f \,\frac{\partial v}{\partial x_i} \,\mathrm{d}^N x$$

exists for arbitrary test functions $v \in C_c^\infty(\Omega)$, the function g_i is called the first order weak partial derivative with respect to x_i of f in Ω and denoted either by $D_i f$ or just by $\frac{\partial f}{\partial x_i}$ or $f_{,i}$, as long as the meaning is clear from the context.

Defining higher order weak partial derivatives involves mixed partial derivatives which makes notation more cumbersome. Since, these derivatives are not required in the main body of the book, their definition is omitted here.

Based on weak first-order partial derivatives, corresponding Sobolev spaces can be defined in combination with L^p-norms in $\mathbb{R}^N$. According to the requirements of the main part of the book, only the case $p = 2$ is considered here.

Definition C.18 (*The space* $W^{1,2}(\Omega)$, $\Omega \subset \mathbb{R}^N$) The space of all real functions f over $\Omega \subset \mathbb{R}^N$ with finite L^2 norm

$$||f||_{1,2} = \left(\int_\Omega f^2 \, \mathrm{d}^N x + \left[\sum_{j=1}^N \int_\Omega (D_j f)^2 \, \mathrm{d}^N x \right] \right)^{\frac{1}{2}}$$

is called the Sobolev space $W^{1,2}(\Omega)$, for $\Omega \subset \mathbb{R}^N$.

The L^2 norm in $\mathbb{R}^N$ is an inner product. Therefore, $W^{1,2}(\Omega)$ introduced by means of Definition C.18 is a Hilbert space, and as such often just called $H^1(\Omega)$. A notational distinction with the $H^1(\Omega)$ based on Definition C.15 is not really necessary because it is in general clear from the context, whether Definitions C.15 or C.18 applies.

For $\Omega \subset \mathbb{R}^N$, boundary conditions require additional effort due to the following reason. Functions belonging, for instance, to $H^1(\Omega)$ are equivalent if they differ only on a set of measure zero, because the norm in $H^1(\Omega)$ is defined in terms of the Lebesgue integral (Appendix B, Sect. B.8). Since the boundary $\partial\Omega$ is actually a set of measure zero for the norm used in Ω, it is not immediately clear how to establish boundary conditions correctly.

For $\Omega \subset \mathbb{R}$, it can be shown, that every equivalence class of $H^1(\Omega)$ has exactly one continuous representative. Therefore, prescribing values of a function which belongs to $H^1(\Omega)$ at the boundary points is no problem at al if $\Omega \subset \mathbb{R}$.

The situation changes drastically for higher dimensions, i.e., $N \geq 2$. However, if the boundary $\partial\Omega$ of a domain $\Omega \subset \mathbb{R}^N$ is for instance Lipschitz, the trace theorem ensures the existence of a continuous mapping

$$\gamma : H^1(\Omega) \rightarrow L^2(\partial\Omega) \tag{C.15}$$

if

$$||\gamma(u)||_{L^2(\partial\Omega)} \leq K||u||_{H^1(\Omega)} \tag{C.16}$$

holds for all $u \in H^1(\Omega)$ where K is a positive constant. In order for a boundary $\partial\Omega$ to be Lipschitz, it has to be possible to map $\partial\Omega$ to a Lipschitz continuous function. For a definition of Lipschitz continuity, see Definition B.15. The mapping γ is known as trace operator.

Boundary value problems the solutions of which are vector fields, also called vector valued functions, introduce new challenges. Defining proper function spaces requires, first of all, an extension of Lebesgue's integration theory, since the latter has been developed for the case $f : \Omega \rightarrow \mathbb{R}$ with $\Omega \subset \mathbb{R}^N$. Now, however, points of $\mathbb{R}^N$ are no longer mapped to real numbers but to elements of a vector space, for instance, $\boldsymbol{f} : \Omega \rightarrow \mathbb{R}^N$. A theory providing the necessary tools is known as Bochner's integration theory, which even considers general Banach spaces as co-domains, see, e.g., [1]. The corresponding Sobolev spaces are usually denoted by $H^k(\Omega; \mathbb{R}^N)$ or $(H^k(\Omega))^N$. For instance, the appropriate space for Linear Elastostatics is $H^1(\Omega; \mathbb{R}^3)$ because only first order partial derivatives appear in the variational form and the displacement field is a vector field of dimension three.

The general procedure, however, does not change at all. Again, a variational formulation

$$a(\boldsymbol{u}, \boldsymbol{v}) = b(\boldsymbol{v})$$

is required, but now for the trial and test vector fields $\boldsymbol{u}$ and $\boldsymbol{v}$, respectively. In the best case scenario, a single function space V can be defined for trial and test vector fields such that $a : V \times V \rightarrow \mathbb{R}$ and $b : V \rightarrow \mathbb{R}$. Under these conditions, the Lax-Milgram Lemma can be used to discuss existence and uniqueness of the solution, which now requires Korn's inequlity for showing that $a(\boldsymbol{u}, \boldsymbol{v})$ is coercive. For details, see, e.g., [2].

C.10 Bibliographical Remarks

As introductory texts for engineers, we recommend [3, 4]. More advanced but highly recommendable sources are [5, 6] in addition to classics such as [7, 8]. Highly recommendable is as well [2] since it applies methods of Functional Analysis directly to problems of Continuum Mechanics of solids.

References

1. J. Diestel, J. Uhl, *Vector Measures*. Mathematical surveys and monographs (American Mathematical Society, 1977)
2. M. Kružík, T. Roubíček, *Mathematical Methods in Continuum Mechanics of Solids*. Interaction of Mechanics and Mathematics (Springer International Publishing, 2019)
3. H. Shima, *Functional Analysis for Physics and Engineering* (CRC Press, London, England, 2016)
4. B. Reddy, *Introductory Functional Analysis: With Applications to Boundary Value Problems and Finite Elements*. Introductory Functional Analysis Series (Springer, Berlin, 1998)

5. H. Le Dret, B. Lucquin, *Partial Differential Equations: Modeling, Analysis and Numerical Approximation*, 1st edn. International Series of Numerical Mathematics (Birkhauser, Basel, Switzerland, 2016)
6. J. Mason, in *Methods of Functional Analysis for Application in Solid Mechanics*, *Studies in Applied Mechanics*, vol. 9 (Elsevier, 1985), pp. 85–123
7. S. Kesavan, *Topics in Functional Analysis and Applications* (Wiley, New York, 1989)
8. S.C. Brenner, R. Scott, *The Mathematical Theory of Finite Eement Methods*, 3rd edn. Texts in applied mathematics (Springer, New York, NY, 2007)

Appendix D
Solutions of Selected Problems

In the following, parts of the solutions for selected exercises are given. The full solution is only given for a limited number of exercises.

2.1 The solution for u reads

$$u(x) = \frac{P_l\, l}{\lambda_0} \begin{cases} \frac{x}{l} & 0 \le x < a \\ \frac{x+a[\beta-1]}{\beta\, l} & a \le x \le l \end{cases} .$$

2.3 The weak form reads

$$\frac{\lambda_0\, l}{n_0} \int_0^l u'\, v\, \mathrm{d}x = \int_0^{l/2} x\, v\, \mathrm{d}x + \frac{l}{2} \int_{l/2}^{l} v\, \mathrm{d}x \, .$$

We recast the terms of the right hand side properly and apply integration by parts which gives

$$\begin{aligned} \frac{\lambda_0\, l}{n_0} \int_0^l u'\, v\, \mathrm{d}x = & \int_0^{l/2} \left[\frac{x^2}{2} + d_1 \right]' v\, \mathrm{d}x + \frac{l}{2} \int_{l/2}^{l} [x + d_2]' v\, \mathrm{d}x \\ = & - \int_0^{l/2} \left[\frac{x^2}{2} + d_1 \right] v'\, \mathrm{d}x - \frac{l}{2} \int_{l/2}^{l} [x + d_2] v'\, \mathrm{d}x \\ & + v \left(\frac{l}{2} \right) \left[d_1 - d_2 \frac{l}{2} + \frac{l^2}{8} \right] , \end{aligned}$$

taking into account, that the test function v vanishes at $x = 0$ and $x = l$. The equation can only be fulfilled if the boundary terms vanish, which establishes a relation between d_1 and d_2. Furthermore,

U. Mühlich, *Enhanced Introduction to Finite Elements for Engineers*, Solid Mechanics and Its Applications 268, https://doi.org/10.1007/978-3-031-30422-4

$$u'(x) = -\frac{n_0 l}{2\lambda_0}\begin{cases} \left(\frac{x}{l}\right)^2 + d_1\frac{2}{l^2} & 0 < x \le \frac{l}{2} \\ \frac{x}{l} + d_1\frac{1}{l^2} + \frac{1}{8} & \frac{l}{2} < x < l \end{cases}$$

must hold. Direct integration leads to two more integration constant. Hence, there are three remaining constants in total, which can be determined by applying the homogeneous Dirichlet boundary conditions as well as the compatibility condition at $x = l/2$. The final result reads

$$u(x) = \frac{n_0 l^2}{48\lambda_0}\begin{cases} 13\frac{x}{l} - 8\left(\frac{x}{l}\right)^3 & 0 < x \le \frac{l}{2} \\ \left[5 + 12\left(\frac{x}{l}\right)^3\right]\left[1 - \frac{x}{l}\right] & \frac{l}{2} < x < l \end{cases}.$$

It is worth noting, that the solution is not twice continuously differentiable in the classical sense. Therefore, a strong form of this problem does not even exist for the considered interval but only for the sub-intervals $(0, l/2)$ and $(l/2, l)$.

2.8 The stiffness of a planar frame element of length l_e and inclined by an angle α can be written as the sum of two matrices

$$\underline{\underline{K}}_e^{\mathrm{f},\alpha} = \underline{\underline{K}}_e^{\mathrm{t},\alpha} + \underline{\underline{K}}_e^{\mathrm{b},\alpha}.$$

Considering constant values for Young' modulus E, cross section area A and area moment of inertia I, these matrices read

$$\underline{\underline{K}}_e^{\mathrm{t},\alpha} = \frac{EA}{l_e}\begin{bmatrix} a & b & 0 & -a & -b & 0 \\ b & c & 0 & -b & -c & 0 \\ 0 & 0 & 0 & 0 & 0 & 0 \\ -a & -b & 0 & a & b & 0 \\ -b & -c & 0 & b & c & 0 \\ 0 & 0 & 0 & 0 & 0 & 0 \\ 0 & 0 & 0 & 0 & 0 & 0 \end{bmatrix},$$

$$\underline{\underline{K}}_e^{\mathrm{b},\alpha} = \frac{EI}{l_e^3}\begin{bmatrix} 12c & -12b & 6h^2s & -12c & 12b & 6h^2s \\ -12b & 12a & -6h^2c & 12b & -12a & -6h^2c \\ 6h^2s & -6h^2c & 4h^3 & -6h^2s & 6h^2c & 2h^3 \\ -12c & 12b & -6h^2s & 12hc & -12hcs & -6h^2s \\ 12b & -12a & 6h^2c & -12hcs & 12ha & 6h^2c \\ 6h^2s & -6h^2c & 2h^3 & -6h^2s & 6h^2c & 4h^3 \end{bmatrix},$$

with $a = \cos^2\alpha$, $b = \sin\alpha\cos\alpha$, $c = \sin^2\alpha$, $s = \sin\alpha$ and $c = \cos\alpha$.

Index

U. Mühlich, *Enhanced Introduction to Finite Elements for Engineers*, Solid Mechanics and Its Applications 268, https://doi.org/10.1007/978-3-031-30422-4

Printed in the United States
by Baker & Taylor Publisher Services

Printed in the United States
by Baker & Taylor Publisher Services